CALIFORNIA MOUNTAIN WILDFLOWERS

California
Mountain
Wildflowers

by Philip A. Munz

UNIVERSITY OF CALIFORNIA PRESS

BERKELEY, LOS ANGELES, LONDON

UNIVERSITY OF CALIFORNIA PRESS
Berkeley and Los Angeles, California

UNIVERSITY OF CALIFORNIA PRESS, LTD.
London, England

CONTENTS

INTRODUCTION

This book is an attempt to present to the general reader who is not trained in taxonomic botany, but who is interested in nature and in his surroundings, some of the wildflowers of the California mountains in such a way that they can be identified without technical knowledge. These are naturally mostly summer wildflowers, together with a few of the more striking species of spring and autumn. They are roughly those from the Yellow Pine belt and upward through Red Fir and Subalpine forests to the peaks above timber line. Obviously the 276 presented cannot begin to cover all that occur in so great an altitudinal range, especially when the geographical limits of the pine belt are considered.

One's first thought of pine belt in California mountains is naturally of the Sierra Nevada, but of course there are also the southern Cascades (Mounts Shasta and Lassen in California) and the Siskiyou Mountains in the northwestern part of the state. Then as one moves south in the North Coast Ranges through the Yolla Bolly Mountains to Snow Mountain and environs in Lake County, he may even go into the Santa Lucia Mountains of the South Coast Ranges. But the pine belt extends even farther south by following the Sierra Nevada through the Tehachapi Range, Mount Pinos, the San Gabriel, San Bernardino, and San Jacinto mountains, all of which have an extensive area altitudinally and spatially in the pine belt. Even the mountains of San Diego County, such as the Palomar, Cuyamaca, and Laguna ranges, have many wildflowers and trees common in the Sierra Nevada. Perhaps 100 species of the Sierran pine belt reach San Diego County, another 100 find their southern limit in the San Jacinto and Santa Rosa mountains of western Riverside County, and 125 more reach the San Bernardino Mountains. In this book I have included also some plants found only in these southern ranges, as well as some confined to the Coast Ranges. I have made no attempt to represent the higher desert mountains, like the White and Panamint ranges, since for the most part the wildflower seeker does not travel in the desert in the summer.

The California Mountains

In general, the mountains of California consist of two great series of ranges: an outer, the Coast Ranges, and an inner, the Sierra Nevada

1

plus the southern end of the Cascade Range, including Mount Lassen and Mount Shasta. The Sierra Nevada, an immense granitic block 400 miles long and 50 to 80 miles wide, extends from Plumas County to Kern County. It is notable for its display of cirques, moraines, lakes, and glacial valleys and has its highest point at Mount Whitney at 14,495 feet above sea level. The Cascade Mountains on the other hand are volcanic, with many extinct volcanoes, the highest in California being Mount Shasta with an elevation of 14,161 feet. The California Coast Ranges are bounded on the north by the Klamath Mountains, with the Siskiyou Mountains being the best known. The Coast Ranges, several more or less parallel series of outer and inner ranges with intervening valleys, are divided into those north and south of the San Francisco Bay area, the North Coast Ranges and the South Coast Ranges respectively. Southward there are the so-called Transverse Ranges, which are primarily, so far as the pine belt goes, the San Gabriel and the San Bernardino mountains, the latter with the highest point in southern California, namely Mount San Gorgonio (Grayback) at 11,485 feet. Then, in a more northerly and southerly direction, come the Peninsular Ranges (Santa Ana, San Jacinto, Santa Rosa, Palomar, Cuyamaca, and Laguna mountains), with Mount San Jacinto at 10,800 feet being the highest.

For the most part the pine belt receives considerable precipitation, usually above 25 inches a year, a large proportion of it as snow. In the Sierra Nevada the snowfall may be tremendous, even up to about 450 inches. Freezing temperatures thus are usual for several months in the year, but summers may be quite warm during the day and with dry air and intense sun. While the California mountains have some summer rain, it is not nearly so great as that of the Rocky Mountains, for instance. It is not surprising then, that in many ways the Sierran flora is not as rich as that of some other great mountain ranges of the same latitude and that, although it has sufficient elevation and winter for circumpolar plants, their number is not nearly so large as in the Rocky Mountains. Much of the montane flora of the California higher mountains has evolved from local sources and local groups of plants, instead of being related to plants of more northern localities. There are of course some circumpolar or far northern species, as for examples: *Juncus bufonius, Festuca brachyphylla, Populus tremuloides, Oxyria digyna, Stellaria longipes, Saxifraga Mertensiana, Mimulus guttatus,* and others. But many montane species are quite local, some even being confined to a single mountain range or even part of that range, as are *Delphinium polycladon* and *Sidalcea reptans* of the Sierra Nevada; *Sidalcea pedata, Arabis Parishii,* and *Castilleja cinerea* of the San Bernardino Mountains; *Draba pterospora* of the Marble Mountains, Siskiyou County; *Arabis rigidissima* of the Trinity Mountains in Trinity and Humboldt counties; and *Leptodactylon Jaegeri* of the San Jacinto Mountains in Riverside County.

One of the interesting features of our mountains is that the dry summer makes possible a great differentiation of habitats. There are the moist often grassy streambanks, wet meadows and even swampy places, ponds with sandy or muddy shores, dryish flats either exposed or shaded by trees, dry rocky slopes and points, talus slopes of loose masses of small stones, and even sheer walls of rock with mere crevices for plant growth. All these habitats can and do have different plants even when at the same elevation. Then too the change in altitude alone makes for very different climate and length of growing season, with marked effect on the species to be found.

For the most part this book has considered the vegetation of the Yellow Pine Forest and above. This Forest begins mostly at an elevation of 2,000 to 3,000 feet in the northern parts of the state and runs up to 6,000 or 7,000 feet, while in the southern counties it ranges from about 5,000 to 8,000 feet. It is characterized by Yellow Pine, Sugar Pine, Douglas-Fir, Incense-Cedar, White Fir, and Black Oak, and has a growing season of four to seven months. Above it is found Red Fir Forest, at 5,500 to 7,500 or 9,000 feet in northern California and 8,000 to 9,500 feet in the south. Its growing season is three to four and one-half months and its characteristic trees are Red Fir, Jeffrey Pine, Western White Pine, Chinquapin, and Quaking Aspen. Next higher comes Lodgepole Forest, from about 8,300 to 9,500 feet and found largely north of the central Sierra Nevada. The growing season is nine to fourteen weeks and the dominant trees are Lodgepole or Tamarack Pine and Hemlock. Above 9,500 feet and poorly represented in southern California and above 8,000 or 9,000 feet in the more northern parts of the state is Subalpine Forest, our most boreal forest in character. The growing season is only about seven to nine weeks and killing frost is possible in every month. Characteristic trees are Limber Pine, Foxtail Pine, Whitebark Pine, Lodgepole Pine, and Mountain Hemlock. Next altitudinally come Alpine Fell-fields or the area above timber line, above 11,500 feet in the Sierra Nevada and 9,500 feet in the North Coast Ranges. Here is found little else except perennial herbs, some of them rather woody. They are scattered or form a low turf or grow among rocks. Many of them form cushions.

HOW TO IDENTIFY A WILDFLOWER

To refresh the reader's memory a drawing is given in figure A showing the principal parts of a typical flower. In the text of the book it seems impossible to discuss plants and their flowers without using the names of some of the parts. In the typical flower we begin at the outside with the *sepals*, which are usually green, although they may be colored. The *sepals* together constitute the *calyx*. Next comes the corolla made up of separate petals, or the petals may be grown together to form a tubular or bell-shaped or wheel-shaped corolla. Usually the corolla is the conspicuous part of the flower, but it may be reduced or lacking altogether (as in grasses and sedges) and its function of attraction of insects and other pollinators may be assumed by the calyx. The calyx and corolla together are sometimes called the perianth, particularly where they are more or less alike. Next as

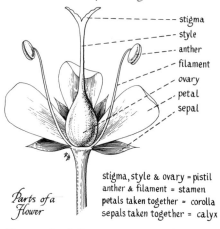

stigma
style
anther
filament
ovary
petal
sepal

stigma, style & ovary = pistil
anther & filament = stamen
petals taken together = corolla
sepals taken together = calyx

Parts of a Flower

FIGURE A. A REPRESENTATIVE FLOWER

we proceed inward in the flower, we usually find the *stamens*, each consisting of an elongate *filament* and a terminal *anther*, in which latter pollen is formed. At the center of the flower are one or more *pistils*, each with a basal *ovary* containing the ovules or immature seeds, a more or less elongate *style*, and a terminal *stigma* with a rough sticky surface for catching pollen. In some species, stamens and pistil are borne in separate flowers or even on separate plants. In the long evolutionary process by which plants have developed into the many diverse types of the present day and by which they have been adapted to different pollinating agents, their flowers have undergone very great modifications and so now we find more variation in the flower than in any other plant part. Hence classification is largely dependent on the flower.

To help the reader identify a wildflower, color plates are given for 96 species and the drawings where used are grouped by color. In attempting to arrange plants by flower color, however, it is difficult to place a given species to the satisfaction of everyone. The range of color may vary so completely from deep red into purple, or from white to whitish to pinkish, or from blue into lavender, that it is impossible to satisfy the writer himself as to whether one color group or another should be used, let alone his readers. I have done the best I could and have tried for the general impression given as to color, especially when the flowers are minute and the general color effect may be caused by other parts than the petals. My hope is that by comparing a given wildflower with the drawing it resembles under the color the reader thinks he would place it and then by checking with the facts given in the text, he may in most cases succeed in discovering what his plant may be.

Which Wildflowers Are to be Sought in This Book

One of the big problems that the writer of such a book faces is which species to include in it. Of course the title suggests a given limitation: plants of the pine belt and above. In my book *California Spring Wild-flowers* I included many things from the Redwood and other coastal areas and from below the pine belt, most of them being spring bloomers. In *California Desert Wildflowers* are to be sought the plants from below the pine belt and on the desert slopes. So, in the present volume we look for the truly montane species, which are primarily summer bloomers. But here I can present only 276 species out of the thousand or so that grow in the area. If all were illustrated, the volume would be unwieldy and expensive. I have tried therefore to select plants that are representative of their groups and in many cases to mention others in the text. I have presented not only species that are showy and naturally get attention, like the lilies, but others that are unusual in one way or another and arouse the finder's curiosity.

I have tried not to be too local in selection of species in order that the book may be of use in various parts of the state rather than in just the Sierra Nevada, although naturally that area is the most representative of the California mountains.

The scientific nomenclature used in this volume is that employed in the more complete and technical work by Munz and Keck, *A California Flora,* University of California Press, 1959. When more information is needed than is available in this wildflower book, reference to the larger volume is recommended.

Acknowledgments

Most of the drawings in this book were made by Jeanne R. Janish (Mrs. Carl F. Janish), whose illustrations in many books on western plants are well known and who has an unusual ability to capture the living appearance of a species even when working largely with pressed specimens. A few by Tom Craig and Rodney Cross were made for my 1935 *Manual of Southern California Botany,* which has been out of print for many years. The kodachromes belong largely to the collection of the Rancho Santa Ana Botanic Garden, many of them having been taken by Percy C. Everett of the Garden staff, by Stephen S. Tillett, and by Harry L. Buckalew of Fresno, California. Two (*Sarcodes* and *Cassiope*) were by Brooking Tatum. Others were loaned to me by Mr. Buckalew from his private collection, namely: *Adiantum pedatum, Pellaea Breweri, Allium validum, Actaea rubra* ssp. *arguta, Delphinium glaucum, Darlingtonia californica, Heuchera rubescens, Spiraea densiflora, Sorbus scopulina, Linum perenne* ssp. *Lewisii, Chimaphila umbellata* var. *occidentalis, Pyrola asarifolia, Castilleja Lemmonii, C. nana, Penstemon heterodoxus, Mimulus cardinalis,* and *Valeriana capitata* ssp. *californica.* The following are the property of the Jepson Herbarium of the University of California and were kindly made available by Dr. Rimo Bacigalupi; of them the first two are by Charles Webber and the others by the late G. Thomas Robbins: *Fritillaria recurva* and *Ceanothus prostratus; Fritillaria pinetorum, Erysimum perenne, Sedum Rosea, Lupinus Breweri, Epilobium angustifolium, E. obcordatum, Primula suffrutescens, Penstemon corymbosus, Mimulus Whitneyi, Sambucus microbotrys,* and *Solidago multiradiata.* Professor Homer Metcalf of Montana State College has loaned me *Smilacina racemosa, Ranunculus Eschscholtzii, R. aquatilis, Lewisia rediviva, Heracleum lanatum, Phlox diffusa, Phacelia curvipes,* and *Pedicularis groenlandica.* Mr. John Olmsted, a former graduate student at Claremont, put these at my disposal: *Ranunculus alismifolius* var. *alismellus, Sedum spathulifolium* and *Aster alpigenus* var. *Andersonii.* To all these gentlemen and to their institutions where such are involved, I am greatly obligated.

It is an especial pleasure to express my appreciation of her great help and wise counsel to Susan J. Haverstick given in editing this book as well as the two preceding on California wildflowers and the larger, more technical *A California Flora.*

Philip A. Munz
Rancho Santa Ana Botanic Garden
Claremont, California
October 1, 1962

FERNS AND FERN ALLIES

Section One

Although not "wildflowers" in any sense at all, ferns and their allies are plants of woods and fields and interest so many of us that I cannot resist putting in a few of the more conspicuous montane forms (see also pages 27–28). COMMON HORSETAIL (*Equisetum arvense*), figure 1, is one of these. Early in the spring the underground rootstocks send up simple, pale, flesh-colored, jointed stems to the height of one foot, having pale sheaths at the nodes with eight to twelve lance-shaped brownish teeth. At the tips of these stems are borne elongate cones about an inch long formed of whorls of stalked peltate structures on the under side of which grow many spore-producing organs (spores being single-celled reproductive bodies). The fertile stems are soon followed by green sterile stems one-half to two feet high and with slender branches in dense whorls so as to appear quite bushy.

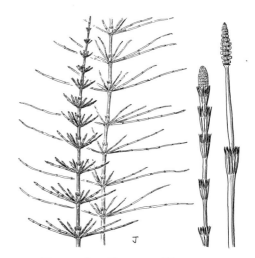

FIGURE 1.   COMMON HORSETAIL

GRAPE FERN (*Botrychium multifidum*), figure 2, is a stout rather fleshy fern to almost two feet high with a large, triangular, much-divided frond from which rises a fertile "spike" resembling a cluster of grapes. Its ultimate divisions bear roundish sporangia that produce the spores. This fern occurs in moist places and borders of woods from about 4,000 to 11,200 feet from Monterey and San Bernardino counties northward.

FIGURE 2.   GRAPE FERN

BRACKEN or BRAKE FERN (*Pteridium aquilinum* var. *lanuginosum*), figure 3, is a coarse fern with long-creeping branched underground parts and erect or ascending green leaves up to several feet high. The blade or expanded part of the leaf is much divided, triangular to elongate, with the margins of the fertile parts curled under to form a protected area where the spores are produced. In

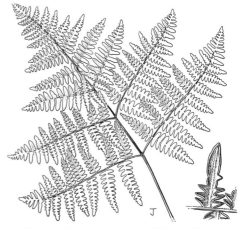

FIGURE 3.   BRACKEN or BRAKE FERN

FIGURE 4. LACE FERN

FIGURE 5. BRITTLE FERN

FIGURE 6. LADY FERN

the pine belt Brake is a widely distributed groundcover in the forests and ranges to elevations of about 10,000 feet.

LACE FERN (*Cheilanthes gracillima*), figure 4, is a tufted plant to about one foot tall, the many leaves having dark brown stems and being twice divided into dull or yellowish-green segments that are brownish-hairy beneath. It grows among rocks at 2,500 to 9,000 feet, from Yellow Pine Forest to Lodgepole Forest and from Marin and Tulare counties northward. Somewhat resembling it is *Cheilanthes intertexta* (COASTAL LIP FERN) with the blades divided three to four times and having bright chestnut-brown scales beneath.

BRITTLE FERN (*Cystopteris fragilis*), figure 5, is more delicate than the foregoing ferns. It has few to several fronds arising from the rootstock, with slender, brittle, smoothish stems and more or less elongate blades a few inches to about one foot long. It has thin pointed leaf-segments with minute rounded fruit-dots or fertile spots beneath. It is found mostly in moist rocky places, frequently quite shaded, and up to 12,500 feet high, in most mountains and far afield.

LADY FERN (*Athyrium Filix-femina* var. *californicum*), figure 6, is a rather large robust plant mostly between 4,000 and 9,500 feet. It is tufted from a short, usually erect, stout rootstock, the fronds being two to five feet high, two to three times pinnately divided and with numerous minute fruit-dots beneath that produce the spores. It occurs from the San Bernardino and San Jacinto mountains northward through the Sierra Nevada and in the Coast Ranges from the Yolla Bolly Mountains north. A related form is found at lower altitudes near the coast of central and northern California.

FLOWERS ROSE TO PURPLISH-RED OR BROWN

OR BROWN

Section Two

TIMBERLINE BLUEGRASS (*Poa rupicola*), figure 7, is included here, largely to show what a grass is and to distinguish it from a sedge (figure 8). Grasses in general have hollow stems with a solid swelling or node at the base of each leaf, which latter ensheaths the stem for a distance, then runs out into the elongate blade. The flowers are minute, without calyx or corolla, and in small groups or spikelets, each of these with two bracts or glumes at the base. Timberline Bluegrass is tufted, less than a foot high, with short stiffish inrolled leaves and slender purplish panicles of spikelets. It grows on rocky screes and ridges at 11,000 to 13,-000 feet, from Mono County to Tulare County.

Sedges are exemplified by SIERRA SEDGE (*Carex Helleri*), figure 8, a more or less tufted plant to about one foot high and with terminal clusters of purplish, almost black, bracts subtending the flowers which are encased in saclike scales. It is found on rocky and gravelly slopes at 8,500 to 13,600 feet from Tulare County to Mount Lassen. In the sedge, as compared with the grass, the leaves are three-ranked (not two-), the leaf-sheaths have united margins, and the stem is solid and usually triangular. Each flower has a glume. Both groups are important for grazing and for hay.

WILD ONION (*Allium campanulatum*), figure 9, of the Lily Family (see pages 28, 29, 39, 52, 62, 83, 93), belongs to a large group of California bulbous plants with a strong onion taste and odor when crushed. This species may be to about one foot high and has two to three flat leaves. At the base of the flower cluster are two ovate bracts subtending fifteen to forty loosely arranged, pale rose flowers to about one-third inch long. It is

FIGURE 7.   TIMBERLINE BLUEGRASS

FIGURE 8.   SIERRA SEDGE

FIGURE 9.   WILD ONION

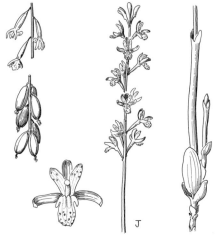

FIGURE 10.   CORAL ROOT

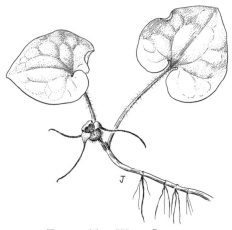

FIGURE 11.   WILD-GINGER

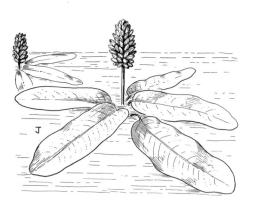

FIGURE 12.   WATER SMARTWEED

found on dry slopes in woods, at 2,000 to 8,500 feet, from the mountains of San Diego County to Oregon, flowering from May to July.

CORAL ROOT (*Corallorhiza maculata*), figure 10, is a saprophytic orchid; that is, it lacks chlorophyll and has to depend for its nourishment on decaying organic matter. The stems are brownish to purplish or yellow, with whitish sheaths, become one to two feet tall, and have scalelike leaves. The flowers are crimson-purple to somewhat greenish, about one-third inch long, and have a white lip spotted with crimson. It is found in montane woods below 9,000 feet from San Diego County to British Columbia and the Atlantic Coast. Flowering is from June to August. See also page 63.

WILD-GINGER (*Asarum Hartwegii*), figure 11, is a stemless perennial herb with creeping rootstocks and with heart-shaped leaves two to three inches long. The brownish-purple flowers have no petals, but the hairy sepals are long-tailed at the tips. When crushed the plant has a spicy odor. Found in shaded places at 2,500 to 7,000 feet, it occurs from Tulare and Trinity counties northward into southern Oregon. The flowers appear in May and June, although almost concealed by the leaves.

WATER SMARTWEED (*Polygonum amphibium* var. *stipulaceum*), figure 12, is an aquatic perennial with floating lance-shaped or lance-oblong leaves with blades two to four inches long. The small rose-colored flowers are in terminal short-cylindrical to ovoid spikes that project from the surface of the water. It occurs in ponds and lakes below 10,000 feet, from San Diego County to Alaska and the Atlantic Coast and blooms from July to September. For other members of the

Buckwheat Family see pages 15, 29, 53, 63, and 95.

PARISH'S WILD BUCKWHEAT (*Eriogonum Parishii*), figure 13, is a montane species of a large group of annuals from lower and middle altitudes. Erect, to about one foot high, diffusely branched into a dense mass of very slender ultimate branchlets that become purplish on aging, it has minute flowers with three outer and three inner pinkish segments. It occurs in dry gravelly places at 4,000 to 9,000 feet, from the southern Sierra Nevada to northern Lower California. The flowering extends from July to September.

FIGURE 13.   PARISH'S WILD BUCKWHEAT

In the Portulaca Family (pages 30 and 65), which is characterized by its two sepals and usually fleshy condition, a conspicuous montane representative is LEWISIA, named for Meriwether Lewis of the Lewis and Clark expedition to the Pacific Northwest. The species shown in figure 14 is *Lewisia Cotyledon* which occurs in a number of forms in the pine belt of Trinity and Siskiyou counties. With a fleshy taproot, it bears several stems to a foot high, many fleshy basal leaves, and many flowers in terminal open clusters. The petals are whitish with red stripes or tinge, half an inch long, and appear in June and July. The plant occurs at 4,000 to 7,500 feet. A closely related species is *L. Cantelowii* found at 3,000 feet in Plumas and Nevada counties.

FIGURE 14.   LEWISIA

SPRING BEAUTY (*Claytonia lanceolata*), figure 15, also of the Portulaca Family, has one to several stems two to six inches high, with some basal leaves and one pair of stem leaves one to two inches long. The pinkish flowers are one-third to one-half inch long and from few to fifteen in number. Spring Beauty occurs

FIGURE 15.   SPRING BEAUTY

FIGURE 16.   BLEEDING HEART

FIGURE 17.   STREPTANTHUS

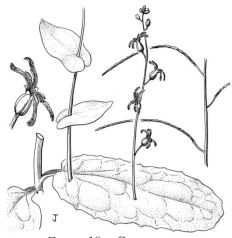

FIGURE 18.   CAULANTHUS

in three different forms, from the San Gabriel Mountains of southern California and the Sierra Nevada to Modoc and Humboldt counties and then to British Columbia and the Rocky Mountains. It flowers from May to July.

BLEEDING HEART (*Dicentra formosa*), figure 16, of the Fumitory Family, has a fleshy rootstock with slender stems to a foot or more high. The leaves are basal, long-petioled, much dissected; the flowers are several, more or less nodding, with generally a rose-purple corolla over half an inch long. The petals are four: the two outer with spreading tips, the two inner wing-crested on the back. The species is found in damp, somewhat shaded places below 7,000 feet, from Santa Cruz and Tulare counties to southern Oregon. Two other species (*D. uniflora* and *D. pauciflora*) with one to three flowers are also montane, at up to 12,000 feet.

STREPTANTHUS (*S. tortuosus*), figure 17, is in the Mustard Family (see pages 17, 54, 67, 68, and 95) with its four sepals, four petals, and biting or acrid taste. Annual or biennial from leafy rosettes, the stems are erect, freely branched above, with clasping stemleaves and terminal racemes of rather purplish flowers. The seed pods are arched-spreading. Several forms occur, the most common in the mountains being the bushy var. *orbiculatus* at 7,000–11,500 feet, or lower farther north. Flowering is primarily from June to September.

CAULANTHUS, a genus closely related to *Streptanthus,* is here represented by S. *amplexicaulis,* figure 18. The slender stem is seven to twelve inches high, with broad leaves to about four inches long. The purple flowers are less than half an inch long and the spreading, curved seed pods about three inches long. It is found on dry loose slopes at 5,000 to 8,500 feet from the San Bernardino Mountains to Mount Pinos in Ventura County, and flowers from May to July.

PHOENICAULIS (*P. cheiranthoides*), figure 19, seems not to have a folk name, an unfortunate circumstance, since the plant is very conspicuous, especially in fruit. A member of the Mustard Family, with a thick perennial stem covered with the bases of dead leaves and with the stem less than a foot long, it produces many pinkish flowers and dense racemes of striking horizontal elongate pods. It is found on dry granitic slopes and benches at 3,800 to 10,700 feet in the Sierra Nevada from Inyo County to Modoc and Siskiyou counties, then south in the Coast Ranges to Mendocino County. Its northward range is to Washington and Idaho.

JAMESIA (*Jamesia americana* var. *californica*), figure 20, is another plant for which I find no common name. A member of the Saxifrage Family, to which the Currant, Mock-Orange, and other familiar shrubs belong (see pages 31, 41, and 68), it is woody, to about three feet high, and deciduous with opposite leaves to about one inch long. The rose-pink flowers are in terminal clusters and one-fourth inch long. Found about rocks at 7,800 to 12,000 feet, Jamesia occurs from the southern Sierra Nevada to Utah and blooms in July and August.

WOOLLY POD (*Astragalus Purshii*), figure 21, is in the large group of Locoweeds of the Pea Family (pages 48, 71, 84, and 98). A perennial, with taproot and tufted or matted stems, it is covered with white wool, even the seed-pods being almost concealed by the shaggy hairs. The flowers are one-half inch or longer and from yellowish with purple

FIGURE 19.   PHOENICAULIS

FIGURE 20.   JAMESIA

FIGURE 21.   WOOLLY POD

FIGURE 22.   CRANESBILL

FIGURE 23.   CHECKER

FIGURE 24.   CALIFORNIA-FUCHSIA

tips to pink or bright purple. The species has several forms in California, but as a group ranges from about 2,000 to 11,000 feet through much of our montane area from San Bernardino Mountains north.

CRANESBILL (*Geranium Richardsonii*), figure 22, is a perennial with one to few stems to about two feet or more high. The lower leaves are long-petioled and deeply 5–7-parted into toothed or cleft segments; the upper have largely three parts. The flowers are scattered, five-petaled, white with reddish or purplish veins. It occurs in moist places as about meadows, at 4,000 to 9,000 feet, from the San Jacinto Mountains to British Columbia. Flowers appear in July and August.

CHECKER (*Sidalcea glaucescens*), figure 23, of the Hollyhock or Mallow Family, has a woody root-crown and slender stems one to two feet long. The leaves are deeply 5–7-lobed or -parted into rather narrow divisions. Flowers are borne in a slender raceme, the pink to rose petals reaching a length of one-third to two-thirds of an inch. This species differs only technically from others of the mountains. It is found in dry grassy places or open woods, at 3,000 to 11,000 feet, from Tulare to Siskiyou and Modoc counties and flowers from May to July.

CALIFORNIA-FUCHSIA (*Zauschneria california* ssp. *latifolia*), figure 24, is a hairy, often somewhat glandular perennial, frequently bushy, the stems being up to two feet long and densely leafy. The leaves are mostly in pairs, ovate to lance-ovate, and often half an inch or more wide. The brilliant red tubular flowers have the four petals characteristic of the Evening-Primrose Family (pages 19, 32, 73, and 100). Widely spread in dry rocky places below 10,000 feet in the montane coniferous forest, California-Fuchsia ranges

from San Diego County to southwestern
Oregon and flowers in late summer.

WILLOW HERB (*Epilobium glaberri-
mum*), figure 25, is a member of the
Evening-Primrose Family with its four
petals and the ovary beneath the flower.
Like California-Fuchsia it has seeds bearing a tuft of hair. There are many species
about mountain meadows and in other
damp places, this one being a perennial
with scaly wiry rootstocks and with
glaucous (whitened with a bloom) upper
parts one to two feet high. The petals
are to about one-fourth inch long,
notched, lavender to pink. Found at
elevations of 3,000 to 11,500 feet, it
ranges from the San Jacinto Mountains
north to Washington and flowers in July
and August.

The MILKWEED genus *Asclepias* is
quite large in California, with mostly
opposite or whorled leaves, milky sap,
and flowers in clusters lacking a central
axis (umbels). *Asclepias Solanoana*,
figure 26, has prostrate stems to about
one foot long, leaves from one to almost
two inches long, and flowers purple outside and white inside, the lobes being
reflexed. It occurs on dry serpentine out-
croppings at 2,000 to 5,000 feet in the
North Coast Ranges from Lake County
to Trinity County, blooming in June.

COLLOMIA (*C. linearis*), figure 27, is
in the Phlox Family (see pages 20, 35,
45, 50, and 76) and belongs to a small
group of herbs with alternate leaves and
funnelform or salverform flowers. This
species is annual, up to about two feet
tall, with the flowers compacted into
bracteate heads. The corolla is pink to
purplish, about one-half inch long. The
plant is found in dry places, generally
between 3,000 and 10,500 feet, from the
San Bernardino Mountains to Alaska and

FIGURE 25.   WILLOW HERB

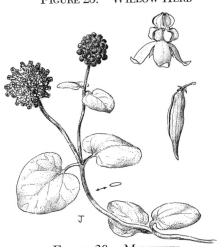

FIGURE 26.   MILKWEED

FIGURE 27.   COLLOMIA

FIGURE 28. PHLOX

FIGURE 29. GILIA

FIGURE 30. BLUE BALLS

Quebec. In California flowering is from May to August.

PHLOX (*P. speciosa* ssp. *occidentalis*), figure 28, has the stems somewhat woody at the base, to a foot or more high, with pairs of narrow leaves. The flowers are few, toward the end of the stems, bright pink, about one-half inch long. This Phlox grows on rocky hillsides and wooded slopes at 1,500 to 7,000 feet, from Fresno and Sonoma counties north and flowers from April to June. The genus *Phlox* is characterized by having the stamens arise at different levels in the corolla tube; for another species see page 45.

GILIA (*G. splendens*), figure 29, has the stamens arise at one level. This species of the Phlox Family is an annual, mostly one to three feet high, with a rosette of basal leaves that are deeply cut into segments. The flowers are in an open terminal inflorescence, pinkish to pinkish-violet, funnelform, one-half to almost an inch in diameter. The species is quite variable and has a number of named forms. It occurs at from 1,000 to 7,000 feet, from the San Jacinto and San Bernardino mountains to those of Monterey County. Flowers appear largely from May to July. See also pages 35 and 77.

BLUE BALLS (*Nama Rothrockii*), figure 30, of the Waterleaf Family (pages 50, 77, 86, and 87), with flowers in coiling or subcapitate clusters or cymes, is a perennial with slender running underground rootstocks and coarsely toothed leaves. The purplish-lavender flowers are funnelform and about one-half inch long. The species is found on dry sandy flats and benches at 7,000 to 10,000 feet from Mono and Fresno counties to Inyo and Tulare counties and in the San Bernardino Mountains. Figure 30 shows also *Nama Lobbii* with entire leaves, a species from Trinity and Siskiyou counties to Nevada County.

Related to the preceding *Nama* is TURRICULA (*T. Parryi*), figure 31, a very glandular, ill-scented, somewhat woody, erect perennial, branched from the base and with very numerous leaves. It grows to a height of three to eight feet and produces numerous purplish funnelform flowers half an inch or more long. It is occasional in dry places, particularly after fire, at 1,000 to 8,000 feet and is found in Fresno and Kern counties of the Sierra Nevada, in the inner Coast Ranges of San Luis Obispo County, and in the mountains of southern California. It produces severe dermatitis on many persons.

GIANT HYSSOP (*Agastache urticifolia*), figure 32, is a tall perennial herb with several stems branched above and with broad toothed leaves. The flowers are rose or violet, about one-half inch long, in dense sessile whorls that may make a more or less continuous spike. This species grows in moist places below 9,000 feet, from the San Jacinto and San Bernardino mountains to British Columbia. It flowers from June to August.

HEDGE-NETTLE (*Stachys rigida*), figure 33, is a hairy perennial with square stems two to three feet high and more or less toothed leaves. The flowers are rose-purple or veined with purple, one-half inch or more long, the upper lip being much shorter than the lower. It runs through a number of forms and ranges from 1,000 to 8,000 feet and from San Diego County to Washington. Flowering is in July and August. The last two plants named above are of the Mint Family (pages 50, 51, and 88).

FIGURE 31.   TURRICULA

FIGURE 32.   GIANT HYSSOP

FIGURE 33.   HEDGE-NETTLE

FIGURE 34.   MOUNTAIN-PENNYROYAL

FIGURE 35.   BIRD'S BEAK

FIGURE 36.   ELEPHANT HEADS

MOUNTAIN-PENNYROYAL (*Monardella odoratissima*), figure 34, is a branched perennial, somewhat woody at the base, the several stems to about one foot long and the leaves to an inch long and very aromatic when crushed. The pale purple flowers, slightly more than one-half inch long, are crowded into heads subtended by colored bracts. It is a member of the Mint Family (pages 21, 51, 88), hence with paired aromatic leaves and two-lipped corollas. The species has several forms and ranges from 3,000 to 11,000 feet from the San Jacinto and San Gabriel mountains through the Sierra Nevada and North Coast Ranges to Oregon, blooming from June to September.

BIRD'S BEAK (*Cordylanthus Nevinii*), figure 35, is a slender-stemmed openly branched annual with yellow roots and linear leaves, the lower having three divisions. The flowers are in heads of one to three, purplish, about one-half inch long, two-lipped, the upper lip dark with white lateral margins. It is found on dry slopes at 5,000 to 8,000 feet, from the San Gabriel Mountains of Los Angeles County to the mountains of San Diego County. Similar species occur in northern and central California.

ELEPHANT HEADS (*Pedicularis groenlandica*), figure 36, like Bird's Beak, belongs to the Figwort Family (pages 35–38, 51, 56, 88, 102), with its two-lipped corollas, but is not usually aromatic as are the Mints. It is a perennial, one to two feet tall, the leaves on the lower parts and saw-toothed. The flowers are borne along the upper stem, red-purple, the upper lip elongated into a slender beak one-sixth to one-fourth inch long. It is occasional in meadows and wet places at 6,000 to 11,200 feet, in the Sierra Nevada and north to boreal

America and to the Atlantic Coast. It flowers from June to August. See especially page 38.

KELLOGGIA (*K. galioides*), figure 37, is a slender perennial herb with creeping rootstocks and several stems to almost one foot high. The paired leaves are not more than an inch long. The corolla is funnelform, small, pinkish, mostly four-lobed. Although not a showy plant, Kelloggia is common on dry benches and slopes at 3,000 to 9,600 feet, from the San Jacinto and San Bernardino mountains northward to Washington and Idaho. Flowering is from May to August. This species belongs to the Madder Family to which the whorled-leaved Bedstraw belongs.

SNOWBERRY (*Symphoricarpos acutus*), figure 38, is a prostrate or trailing shrub with soft-hairy young twigs and paired somewhat-lobed leaves. The bright pink bell-shaped flower is less than one-fourth of an inch long and hairy within. The species is found in rather damp places at 3,500 to 8,000 feet, from Tulare and Lake counties north to Oregon. It belongs to the Honeysuckle Family (see page 46).

TWIN FLOWER (*Linnaea borealis* ssp. *longiflora*), figure 39, another member of the Honeysuckle Family, is named for Linnaeus and his portraits always show him holding a sprig of it. It is a creeping vinelike perennial with paired leaves and two pinkish, funnelform, nodding flowers at the summit of each erect flower-stem. Found in dense woods at 400 to 8,000 feet, this subspecies occurs from Humboldt and Trinity counties to Del Norte and Modoc counties, thence to Idaho and Alaska. The species as a whole is circumpolar.

ALPINE DAISY (*Erigeron vagus*), fig-

FIGURE 37.   KELLOGGIA

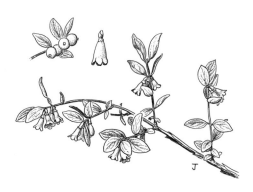

FIGURE 38.   SNOWBERRY

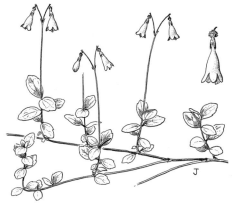

FIGURE 39.   TWIN FLOWER

FIGURE 40.   ALPINE DAISY

FIGURE 41.   EUPATORIUM

FIGURE 42.   THISTLE

ure 40, is a member of a large genus belonging to the Sunflower Family (pages 47, 51, 57, 79, 90, 102–108), in which what looks like a flower is really a head of many small flowers or florets, some of which outer ones (ray-flowers or rays) may simulate petals while the inner ones (disk-flowers) are simply typical minute tubular flowers. This species is a small perennial with a heavy taproot, crowded basal leaves, and purple-tinged rays. It is found in scree and rock-crevices at 11,000 to 14,100 feet from Tulare County to Mono and Tuolumne counties and to Oregon and Colorado. It closely resembles *Aster Peirsonii* of similar elevations which has the bracts subtending the flower head in more numerous overlapping series.

EUPATORIUM (*E. occidentale*), figure 41, is also of the Sunflower Family, but without ray-flowers. A more or less woody tufted perennial, it has mostly alternate leaves with rather compact clusters of flower heads at the ends of the branches. The flowers are pink, red-purple, or whitish. Found about rocks at 6,500 to 11,000 feet in the Sierra Nevada, it ranges at somewhat lower elevations in the Coast Ranges from Tehama County north to Washington and east to Utah.

THISTLE (*Cirsium Andersonii*), figure 42, is a biennial or short-lived perennial, with purplish-red slender stem one to three feet high and often loosely-woolly leaves that have very spiny teeth and lobes. The flower heads (Sunflower Family) have no rays, but the rose-purple disk-flowers are very slender and make showy heads. It is found on dry slopes at 4,000 to 10,500 feet, from Tulare County to Siskiyou and Trinity counties.

COLOR PLATES

A so called Fern Ally (see also page 9) is Little Club-Moss or Spike-Moss (*Selaginella Watsonii*), plate 1, a locally frequent inhabitant of dry rocky places between 7,500 and 14,000 feet, from the Santa Rosa and San Jacinto mountains of southern California, north through the Sierra Nevada to Nevada County and to Oregon and Montana. It has branching prostrate stems one to three inches long with closely overlapping scalelike leaves and terminal erect green "cones" or strobili, in the axils of whose bracts are spore-bearing structures. The leaves have minute bristles or setae along their margins.

PLATE 1.   LITTLE CLUB-MOSS

A true Fern (see also pages 9,10) is Cliff-Brake (*Pellaea Breweri*), plate 2, with short-creeping rhizomes covered with twisted brown scales. The fronds are tufted, two to eight inches high, pinnate into thickish mostly two-parted divisons that are mitten-shaped and with the fruit-dots beneath almost covered by the reflexed leaf-margin. It grows on exposed dry rocky places at 7,000 to 12,000 feet, in the San Bernardino Mountains and from Tulare County to Siskiyou County, thence to Washington and Wyoming. Much like it is *Pellaea Bridgesii* also of the Sierra Nevada, but with the leaf-divisions (pinnae) entire.

PLATE 2.   CLIFF-BRAKE

Rock-Brake or Parsley Fern (*Cryptogramma acrostichoides*), plate 3, is one of the most conspicuous ferns at high elevations, growing in rocky places at 6,000 to 11,000 feet, in the San Jacinto and San Bernardino mountains of southern California, in the Sierra Nevada, and to Alaska and Labrador. It is tufted, to almost a foot high, and has two kinds of leaves, the fertile or spore-bearing fronds having narrower and longer divisions (pinnules) than the sterile, in which latter they are flat and expanded.

PLATE 3.   ROCK-BRAKE, PARSLEY FERN

27

PLATE 4.   FIVE-FINGER FERN

FIVE-FINGER FERN (*Adiantum pedatum* var. *aleuticum*), plate 4, has erect fronds one to two and one-half feet tall, the dark stout stipes (leaf-stems) being forked above and each branch bearing on the outer side several spreading pinnate divisions four to sixteen inches long. The ultimate divisions of the frond are numerous with fruit-dots on the under surface of the truncate tips. It grows in moist shaded places from sea level to 10,700 feet, in the San Bernardino and San Gabriel mountains and along the Coast Ranges and Sierra Nevada to Alaska, Utah, and Quebec.

PLATE 5.   SWORD FERN

SWORD FERN (*Polystichum munitum*), plate 5, is a coarse evergreen fern from upright scaly rhizomes and with many once-pinnate fronds two to four feet high. The pinnae or leaf-divisions are lance-shaped, sharply toothed, and with round submarginal fruit-dots on the under surface. It is common, as at least two named subspecies, on canyon slopes below 8,500 feet from the mountains of San Diego County to the Santa Lucia Mountains and in the Sierra Nevada from Tulare County to Plumas County.

PLATE 6.   WILD ONION

WILD ONION (*Allium validum*), plate 6, placed by some botanists in the Lily Family and by others in the Amaryllis Family (see also page 13), is a bulbous green-leaved onion two to three feet high and common in large clumps in wet meadows and on stream banks. It grows at from 4,000 to 11,000 feet in the Sierra Nevada and in the Coast Ranges from Lake County north to British Columbia and Idaho. This onion is of good flavor. The leaves are three to six in number and flattish, while the many rose-colored flowers are about one-third inch long.

Scarlet Fritillary (*Fritillaria recurva*), plate 7, of the Lily Family (pages 13, 28, 39, 52, 62, 83, 93), has a stem one to three feet high and usually eight to ten leaves in two to three whorls, each with two to five leaves. The flowers are nodding, scarlet, checked with yellow within and tinged with purple without. It is found on dry hillsides in brush or woods at 2,000 to 6,000 feet, in the inner Coast Ranges from Lake County north and in the Sierra Nevada of Nevada County; to Oregon and Nevada. Flowering is from March to July.

PLATE 7.   SCARLET FRITILLARY

Another Fritillary (*Fritillaria pinetorum*), plate 8, has a very glaucous stem three to fifteen inches high with twelve to twenty glaucous leaves that are somewhat whorled. The flowers are erect or nearly so, purplish, mottled with greenish-yellow. It is found on somewhat shaded granitic slopes at 6,000 to 10,500 feet from Alpine County to San Bernardino County. A species resembling it is *F. atropurpurea* with a more slender, less inflated stem and without "rice-grain" bulblets at the base. The flowers are nodding and purplish-brown, spotted with yellow and white.

PLATE 8.   FRITILLARY

Mountain Sorrel (*Oxyria digyna*), plate 9, of the Buckwheat Family (pages 14, 15, 30, 53, 63, 95), is remarkable for its wide distribution. Found in rocky places like crevices in cliffs at 8,000 to 13,000 feet in California, it ranges from the San Jacinto and San Bernardino mountains, Sierra Nevada, and Yolla Bolly Mountains north to Arctic America and Eurasia. It is a low perennial with acid juice, roundish leaves, and small whorled flowers in compact panicles. The reddish or greenish sepals are scarcely one-eighth of an inch long.

PLATE 9.   MOUNTAIN SORREL

PLATE 10.   WILD BUCKWHEAT

Another member of the Buckwheat Family and one of the innumerable kinds of WILD BUCKWHEAT itself is *Eriogonum ovalifolium* ssp. *vineum*, plate 10, a dense, almost matted, white-woolly perennial with leaves scarcely one-half inch long and with flowering stems a few inches high. The mostly pinkish or rose flowers are about one-fourth inch long. It is found in dry, usually rocky places at 5,000 to 11,600 feet, in the San Bernardino Mountains and from the Sierra Nevada to Siskiyou County, then to Oregon and Nevada.

PLATE 11.   LEWISIA

LEWISIA (*L. disepala*), plate 11, of the Portulaca Family (see pages 15 and 65), has a short perennial caudex, stems about one inch long, and fleshy somewhat shorter basal leaves. The flowers are solitary, pinkish, one-half inch long or more. It is found in rocky places at about 6,500 to 8,500 feet, on the summits bordering Yosemite Valley in the Sierra Nevada. Flowering is in May and June.

PLATE 12.   BITTERROOT

BITTERROOT (*Lewisia rediviva*), plate 12, is a widespread and variable Lewisia of loose gravelly slopes and rocky places from 2,000 to 9,000 feet and from southern California to British Columbia and the Rocky Mountains. It is the state flower of Montana. The rose to whitish flowers, from one-half to one inch long, appear in early spring and have many petals. They are followed by fleshy narrow leaves one to two inches long which form rosettes on the surface of the ground.

In the Stonecrop Family (see page 55) is WESTERN ROSEROOT (*Sedum Rosea* ssp. *integrifolium*), plate 13, a fleshy perennial from a short scaly rootstock and with several stems two to eight inches high. The leaves are about one-half inch long. The flowers usually have four sepals and four petals, the latter dark purple and one-eighth of an inch long. Found in moist rocky places at 7,500 to 12,500 feet in California, it ranges from Tulare County to Eldorado County and to Alaska, Siberia, and Colorado. It flowers from May to July.

PLATE 13. WESTERN ROSEROOT

ALUMROOT (*Heuchera rubescens*), plate 14, of the Saxifrage Family (see pages 17, 41, 68, 69), is a variable species with different forms in various parts of the Sierra Nevada. As a species, it has a thick caudex with basal roundish leaves and flowering stems to a foot long and ending in many-flowered panicles. The petals are about one-eighth of an inch long. It is found in dry rocky places at 6,000 to 12,000 feet and is replaced in the mountains of southern California by closely related species.

PLATE 14. ALUMROOT

A Wild Currant of more than usual beauty because of its rose to rather deep red flowers is the SIERRA CURRANT (*Ribes nevadense*), plate 15. It is a slender-stemmed deciduous shrub three to six feet high with roundish leaves one to almost three inches broad. The flowers have reddish sepals and white petals. It grows in moist places and along streams at 3,000 to 10,000 feet, from the Palomar Mountains of San Diego County to southern Oregon, blooming from May to July. The blue-black berries have a bloom and are more or less glandular.

PLATE 15. SIERRA CURRANT

31

PLATE 16.　SPIRAEA

PLATE 17.　FIREWEED

PLATE 18.　ROCKFRINGE

SPIRAEA (*S. densiflora*), plate 16, of the Rose Family (pages 41, 42, 69, 70, 96, 97), is a low shrub with leaves to about one inch long and with a flat-topped inflorescence of pink flowers. The petals are only about one-sixteenth of an inch long, five in number, and the stamens rather many. This montane Spiraea inhabits moist rocky places from 5,000 to 11,000 feet or may range lower in the northern counties. It is distributed from Tulare County in the Sierra Nevada and Trinity County in the Coast Ranges north to British Columbia.

In the Evening-Primrose Family (pages 18, 19, 73, 100, 101), one of the most widely ranging and best known plants is FIREWEED (*Epilobium angustifolium*), plate 17. The seeds being distributed by the wind because of their tuft of hairs, the species comes rapidly into burned and disturbed areas of northern forests. It is a perennial from underground rootstocks, grows to be one to seven feet high, and bears long terminal racemes of mostly rose or lilac-purple flowers with four petals. Found generally below 9,000 feet, it ranges from San Diego County northward to Alaska, the Atlantic Coast, and Eurasia.

Another Epilobium is ROCKFRINGE (*E. obcordatum*), plate 18, with decumbent or matted stems and ovate leaves, each branch bearing one to few flowers near its summit. The petals are rose-purple, one-half inch or more long. It is a strikingly beautiful plant growing largely at the base of rocks on dry ridges and flats at 7,000 to 13,000 feet, from Tulare and Inyo counties to Modoc County and to eastern Oregon and into Idaho. It blooms from July to September.

PIPSISSEWA (*Chimaphila umbellata* var. *occidentalis*), plate 19, is of the Winter-green Family (pages 44, 101), and is a low, evergreen, semishrubby perennial with thick shining leaves and rose-pink flowers. Sepals and petals are each five, concave, round. Pipsissewa is found on dry shrubby slopes in woods at between 1,000 and 10,000 feet, in the San Jacinto and San Bernardino mountains, the Sierra Nevada, and the North Coast Ranges. Its distribution is thence to Alaska and Michigan. It flowers from June to August.

PLATE 19.   PIPSISSEWA

The other two plants on this page are also of the Wintergreen Family. The first one, PINK PYROLA (*Pyrola asarifolia*), plate 20, has two named varieties in California, which are here treated as one entity. The plant is an extensively creeping low perennial herb with leathery broad basal leaves and with flowers at the summit of a leafless somewhat scaly stem. The petals are one-fourth to one-third of an inch long, pink to rose-purple, five in number. Found in California in moist shade and woods below 9,000 feet, Pink Pyrola occurs in the San Bernardino Mountains, Sierra Nevada, and from Mendocino County north to Alaska and to the Atlantic Coast.

PLATE 20.   PINK PYROLA

SNOW PLANT (*Sarcodes sanguinea*), plate 21, is a red fleshy saprophyte (lacking chlorophyll and nourished by decaying material in the soil) with many crowded leaves. The flowers are numerous, in a stout spicate terminal raceme, with a bell-shaped corolla having five broad, red, slightly spreading lobes. Growing in thick humus of forest floors at 4,000 to 8,000 feet, Snow Plant occurs in the Santa Rosa and San Jacinto mountains of Riverside County and north through the Sierra Nevada and in the North Coast Ranges to Oregon.

PLATE 21.   SNOW PLANT

PLATE 22. AMERICAN-LAUREL

AMERICAN-LAUREL (*Kalmia polifolia* var. *microphylla*), plate 22, is a low diffusely branched shrub, with opposite leaves having somewhat inrolled margins. The flowers are in small groups, rose-purple, and almost half an inch in diameter. An inhabitant of boggy places and wet meadows at 7,000 to 12,000 feet in the Sierra Nevada and somewhat lower in the North Coast Ranges, it is found from Tulare and Humboldt counties to Alaska and the Rocky Mountains. Flowering is from June to August. This species and the next belong to the Heath Family (pages 45 and 75).

PLATE 23. MOUNTAIN-HEATHER

MOUNTAIN-HEATHER (*Phyllodoce Breweri*), plate 23, is a low, much-branched, evergreen shrub with narrow alternate leaves. The rose-purple to pinkish flowers are open–bell-shaped, about one-third of an inch long and with long-exserted stamens. It is found in rocky, sometimes rather moist places at 6,000 to 12,000 feet, in the San Bernardino Mountains and the Sierra Nevada from Tulare County north to Mount Lassen. It blooms in July and August.

PLATE 24. SHOOTING STAR

SHOOTING STAR (*Dodecatheon Jeffreyi*), plate 24, is in the Primrose Family (page 35). It is a perennial one to two feet high, each stem having three to eighteen flowers. They are reflexed, one-half to one inch long, with a maroon ring at the base and with magenta to lavender lobes. Stamens are usually four. The leaves are basal and two to twenty inches long. It grows in wet places at 2,300 to 10,000 feet from Tulare and Glenn counties north to Alaska. A closely related species (*D. redolens*) with five stamens is found from the San Jacinto Mountains to Fresno and Inyo counties.

34

SIERRA PRIMROSE (*Primula suffrutescens*), plate 25, is a true Primrose, but unlike our garden primroses, it is woody and with branched creeping stems. The flower stems are two to four inches high with terminal umbels of small flowers. These are magenta with yellow throat. Sierra Primrose is frequent under overhanging rocks and about cliffs, mostly at 8,000 to 13,500 feet, in the Sierra Nevada and in Siskiyou and Trinity counties. It blooms in July and August.

PLATE 25.   SIERRA PRIMROSE

MUSTANG-CLOVER (*Linanthus montanus*), plate 26, of the Phlox Family (see pages 19, 20, 45, 50, 76, 77, 86), is an erect annual up to two feet high. The leaves are remote and cleft into five to eleven linear, stiff-hairy or bristly lobes. The somewhat funnelform corolla is about one inch long, lilac-pink or white, with a purple spot on each lobe. This rather lovely species inhabits dry places at 1,000 to 5,000 feet, in the Sierra Nevada from Nevada County south to the Greenhorn Range in Kern County. Its flowers appear from May to August.

PLATE 26.   MUSTANG-CLOVER

SCARLET PENSTEMON (*Penstemon Bridgesii*), plate 27, of the Figwort Family (see pages 22, 36–38, 51, 56, 88, 89, 102), is one to three feet high and woody at the branched base. It is quite glandular in the inflorescence of scarlet to vermilion flowers which are tubular, an inch or more long, and with the anthers rather horseshoe-shaped. Growing on dry slopes at 5,000 to 10,700 feet, it ranges from the mountains of San Diego County to the Sierra Nevada of Alpine County and to Colorado. Often confused with it is another Scarlet Penstemon (*P. labrosus*) which is not glandular in its upper parts and which has divergent anthersacs.

PLATE 27.   SCARLET PENSTEMON

PLATE 28.   BEARD TONGUE

Another Penstemon is a BEARD TONGUE, so-named because the sterile stamen is strongly bearded. It is *Penstemon Grinnellii*, plate 28, a low spreading perennial wth more or less branched stems four to sixteen inches high and with pairs of rather broad, usually toothed leaves. The inflorescence is rather open with mostly flesh-pink or lavender flowers about an inch long. It is to be sought on dry gravelly, generally granitic slopes at 4,500 to 9,500 feet, from the San Gabriel Mountains to the Santa Rosa Mountains.

PLATE 29.   MOUNTAIN PENSTEMON

MOUNTAIN PENSTEMON (*Penstemon Newberryi*), plate 29, is also a BEARD TONGUE, woody below, matted, the creeping or decumbent stems six to twelve inches long. The rose-red corolla is about one inch long, narrow and only slightly dilated at the throat. Growing in rocky and gravelly places at 5,000 to 11,000 feet, it ranges from Mount Shasta through the higher Sierra Nevada to Tulare County and blooms from June to August.

PLATE 30.   BEARD TONGUE

A BEARD TONGUE that is definitely woody is *Penstemon corymbosus*, plate 30, one to one and one-half feet high and densely glandular-pubescent in the inflorescence. The brick-red corolla is an inch or more long, tubular, with the lower lip spreading. The species is found on rocky slopes and cliffs below 5,000 feet, from Del Norte County to Monterey County and in the foothills of the Sierra Nevada from Shasta County to Sutter County. The flowering season is from June to October.

SCARLET MONKEY-FLOWER (*Mimulus cardinalis*), plate 31, of the Figwort Family, is a viscid-hairy perennial with erect or decumbent stems from one to almost three feet long and with opposite, longitudinally veined, saw-toothed leaves. The scarlet, sometimes yellowish, corolla is about two inches long. Frequenting stream banks and moist places below 8,000 feet, Scarlet Monkey-Flower occurs in most of montane California and in adjacent states. It blooms from April to October.

PLATE 31.   SCARLET MONKEY-FLOWER

PAINT-BRUSH or PAINTED-CUP (*Castilleja miniata*), plate 32, of the Figwort Family, has a long tubular corolla with the lips modified, the upper being elongate and enclosing the style and stamens and the lower being shorter and somewhat saccate. The flower is an inch or more long and the calyx-tips and subtending bracts are quite scarlet. This showy species grows along streams and in wet places below 11,000 feet in the mountains from San Diego County north through the Sierra Nevada and in the Coast Ranges from Glenn County north, to British Columbia and the Rocky Mountains.

PLATE 32.   PAINT-BRUSH

Another smaller and less conspicuous PAINT-BRUSH is *Castilleja nana*, plate 33, grayish-hairy and less than three inches high. The corolla is one-half inch long, the upper lip greenish with purple and white margins. The species occurs in dry rocky places at 8,000 to 12,000 feet from Eldorado and Mono counties to Tulare and Inyo counties and blooms in midsummer.

PLATE 33.   PAINT-BRUSH

PLATE 34.   PAINT-BRUSH

A purplish-red PAINT-BRUSH is *Castilleja Lemmonii*, plate 34, a perennial with several simple stems four to eight inches tall and with linear almost undivided leaves. The corolla is three-fourths of an inch long, glandular-hairy on top. The species occurs in moist meadows at 7,000 to 11,000 feet from Fresno and Inyo counties north to Mount Lassen. Much like it is *Castilleja Culbertsonii* with fewer stems and a less hairy corolla, the lower lip of which is included in the calyx, not exserted as in *C. Lemmonii*.

PLATE 35.   LITTLE ELEPHANT'S HEAD

Two prominent Louseworts of high elevations and also of the Figwort Family are *Pedicularis attolens* and *P. groenlandica*. The former, LITTLE ELEPHANT'S HEAD, plate 35, is a perennial six to fourteen inches high, hairless below and white-woolly in the inflorescence. The leaves are basal and on the lower stem, pinnately divided and with toothed segments. The flowers are in a spikelike raceme, lavender or pink, the upper lip of the corolla recurved and narrowed into an upward beak. The plant is common in meadows and moist places at 5,000 to 12,000 feet in the Sierra Nevada and north to Siskiyou and Modoc counties and into Oregon.

PLATE 36.   ELEPHANT HEADS

Pedicularis groenlandica, plate 36, is called ELEPHANT HEADS and is hairless throughout, one to two feet tall. The beak of the corolla is one-fourth to one-half inch long. The species is occasional in meadows and wet places between 6,000 and 11,200 feet, from the Sierra Nevada to Humboldt, Siskiyou, and Modoc counties. It ranges to boreal America and the Atlantic Coast. See also page 22.

In the Lily Family a tall weedy plant poisonous to livestock is CORN-LILY or FALSE-HELLEBORE (*Veratrum californicum*), plate 37. Forming large clumps three to five feet high, it often fills large patches of meadow and stream bank. It grows at elevations up to 11,000 feet and ranges from the mountains of San Diego County north to Washington and to the Rocky Mountains. Other similar species of northern California have green flowers instead of white and drooping panicle-branches (*V. viride*), fringed perianth-segments (*V. fimbriatum*), and woolly ovary (*V. insolitum*).

PLATE 37.   CORN-LILY

FALSE SOLOMON'S-SEAL (*Smilacina racemosa* var. *amplexicaulis*), plate 38, also of the Lily Family (see pages 13, 28, 29, 52, 62, 83, 93), has a stout root-stock, erect stem one to three feet high, and broad mostly clasping leaves. The small white flowers are in a branched inflorescence, while in the related S. *stellata* the fewer flowers are in a simple raceme. Both species are found in shaded woods, often in moist places, mostly below 8,000 feet, through much of California to British Columbia and the Atlantic Coast.

PLATE 38.   FALSE SOLOMON'S-SEAL

BUSH CHINQUAPIN (*Castanopsis sempervirens*), plate 39, belonging with the oaks to the Beech Family, is a rounded shrub two to eight feet high and with oblong, almost toothless leaves one to three inches long. These are golden or rusty-woolly beneath. The staminate flowers are very ill-smelling, arranged in long catkins, and appear in July and August. The pistillate flowers make spiny burs which a year later are about one inch thick. The species occurs in dry rocky places at 2,500 to 11,000 feet from the San Jacinto Mountains to Oregon.

PLATE 39.   BUSH CHINQUAPIN

39

PLATE 40.   CAMPION

CAMPION (*Silene Sargentii*), plate 40, of the Pink Family (page 65), is a plant of rock crevices and similar places at altitudes of 6,500 to 12,000 feet and is found in the Sierra Nevada from Sierra County to Tulare County. It is tufted, glandular, four to five inches high, and has a somewhat tubular calyx and a white corolla with two-lobed petals, each lobe having a small lateral tooth. Flowers appear in July and August.

PLATE 41.   BANEBERRY

In the Buttercup Family (see pages 47, 53, 66, 83) is BANEBERRY (*Actaea rubra* ssp. *arguta*), plate 41, a perennial herb one to two and one-half feet high. The leaves are on the stems and have broad incised and toothed leaflets. The racemes are dense when in flower, but elongate in fruit, the small white petals soon falling away. The fruits are red or white, rather persistent, about one-fourth of an inch long. Baneberry grows in rich moist woods below 10,000 feet, in the San Bernardino Mountains, Sierra Nevada, and in the Coast Ranges from San Luis Obispo County north. Flowering is in May and June.

PLATE 42.   WATER BUTTERCUP

WATER BUTTERCUP (*Ranunculus aquatilis*), plate 42, as it occurs in California has a number of named varieties which do not need to be considered for our purpose. It is an aquatic perennial with submersed stems and dissected submerged leaves and sometimes lobed floating ones. The five petals are white and about one-fourth inch long. The plant is found in ditches, slow streams, and ponds below 10,000 feet and ranges to Alaska, the Atlantic Coast, and Europe. Flowering is from April to July depending on elevation.

One of our lovely montane flowers is GRASS-OF-PARNASSUS (*Parnassia palustris* var. *californica*), plate 43, of the Saxifrage Family (pages 17, 31, 41, 68). Forming small patches of erect white flowers and growing in wet meadows below 7,000 feet in the Coast Ranges and below 11,000 feet in the Sierra Nevada, it ranges from the San Bernardino Mountains to Oregon. The leaves are entire and in a basal tuft and the flowers are solitary and terminal with entire, conspicuously veined petals one-half inch or longer. Two other species found in California have the petals fringed.

PLATE 43. GRASS-OF-PARNASSUS

A shrub in the same Saxifrage Family is MOCK-ORANGE or SYRINGA (*Philadelphus Lewisii*), plate 44. Most of our California plants fall in the subspecies *californicus* with appressed hairs on the veins of the under surface of the leaves and ranging between 1,000 and 4,500 feet from Tulare County north. But plants with leaves hairy over the whole under surface are the subspecies *Gordonianus* and occur from Lake County north. Petals are one-half to three-fourths of an inch long. The odor is very pleasing. See also page 31.

PLATE 44. MOCK-ORANGE or SYRINGA

Another white-flowered shrub is THIMBLEBERRY (*Rubus parviflorus*), plate 45, of the Rose Family (see pages 32, 42, 69, 70, 96, 97). It is deciduous, mostly three to six feet high, without prickles and with the bark shreddy in age. The palmately five-lobed leaves are four to six inches broad. The flowers are one to two inches across, sometimes pink. The "berry" is scarlet, hemispheric. This plant is found in open woods and in canyons below 8,000 feet, from the mountains of San Diego County to Alaska.

PLATE 45. THIMBLEBERRY

41

PLATE 46. MOUNTAIN-ASH

MOUNTAIN-ASH (*Sorbus scopulina*), plate 46, also of the Rose Family, is shrubby rather than arboreous as are many species in this genus. The bark is thick and reddish; the leaves have eleven to thirteen leaflets. The flat-topped inflorescence has eighty to two hundred flowers, each slightly less than half an inch in diameter. The orange to scarlet fruits (much like tiny apples) are about one-third of an inch in diameter. The species is occasional in canyons and on wooded slopes between 4,000 and 9,000 feet from Tulare County to Siskiyou and Modoc counties, thence to British Columbia and the Rocky Mountains.

PLATE 47. DEER BRUSH

CALIFORNIA-LILAC of the Buckthorn Family (pages 49, 72), is often blue, but in *Ceanothus integerrimus*, plate 47, we have a white-flowered shrub that is often called DEER BRUSH. It has several named varieties, but in general is a loosely branched shrub, three to twelve feet tall. The flowers are sometimes bluish or pink. It is found on dry slopes and ridges between 1,000 and 7,000 feet from southern California to Washington.

PLATE 48. OREONANA

In the Carrot Family (see pages 43, 56, 73, 74), the high mountains have some very interesting plants. Among these are two or three small stemless alpines, like OREONANA (*O. Clementis*), plate 48, of dry granitic gravels at 8,000 to 13,000 feet, in the southern Sierra Nevada. The leaves are decompound, gray-pubescent, an inch or so in length. The flowers and fruit are in globose clusters scarcely exserted from the tuft of leaves, flowering being from late May to August.

One of the larger California members of the Carrot Family is Cow-Parsnip (*Heracleum lanatum*), plate 49, an erect stout plant with broad leaves to as much as twenty inches across. The large compound umbel (flower cluster with its divisions all arising at one point) has fifteen to thirty rays, each bearing a smaller umbel of small white flowers. Cow-Parsnip grows in moist places below 9,000 feet in the San Jacinto and San Bernardino mountains, the Sierra Nevada, and from Monterey County north in the Coast Ranges, to Alaska, and to the Atlantic Coast.

PLATE 49.   Cow-Parsnip

Another large member of the Carrot Family is White Heads or Ranger's Button (*Sphenosciadium capitellatum*), plate 50. It is two to five feet high, with leaves divided into narrow or somewhat wider leaflets. The rays of the umbel bear white heads of tiny flowers. This species grows in swampy places at 3,000 to 10,400 feet, in the San Jacinto and San Bernardino mountains and north through the Sierra Nevada to Mendocino, Siskiyou, and Modoc counties, then to Oregon and Idaho. It flowers in July and August.

PLATE 50.   White Heads

Mountain Dogwood (*Cornus Nuttallii*), plate 51, is a deciduous arborescent shrub or small tree of the Dogwood Family. The leaf-blades are commonly two to four inches long. The small flowers are in a head subtended by large whitish or even pinkish persistent bracts which are quite petaloid and usually appear in the spring before the leaves. Dogwood is found in mountain woods below 6,000 feet, locally in the mountains from San Diego County to Los Angeles County and commonly from Monterey and Tulare counties north to British Columbia.

PLATE 51.   Mountain Dogwood

43

PLATE 52.   PINEDROPS

In the Wintergreen Family (see pages 33, 101) is a group of plants lacking chlorophyll and getting their nourishment from dead or living organisms, as saprophytes and parasites respectively. A parasite on living root-fungi is PINE-DROPS (*Pterospora andromedea*), plate 52, one to three feet tall and with scattered purple-brown scalelike leaves and urn-shaped flowers in a long terminal raceme. It is found in humus in forests at 2,600 to 8,500 feet, from the San Jacinto Mountains to the Sierra Nevada and in the Coast Ranges from Lake County north, to British Columbia and the Atlantic Coast.

PLATE 53.   PITYOPUS

In the same family is PITYOPUS (*P. californicus*), plate 53, a rare plant of deep shade at 1,000 to 5,000 feet. It is reported from Marin, Humboldt, Del Norte, Mendocino, and Fresno counties and from Oregon. It is a waxy-white saprophyte, two to eight inches high, with its white flowers in a dense terminal spikelike raceme.

In the Heath Family (see pages 34, 45, 75) is WESTERN AZALEA (*Rhododendron occidentale*), plate 54, a deciduous shrub one to ten or more feet tall and with shredding bark. The flowers are white to pinkish, the upper lobes with a yellowish blotch, or sometimes there are other color variations. This fragrant Azalea grows in moist places below 7,500 feet, from the Cuyamaca Mountains to the San Jacinto Mountains in southern California, in the Sierra Nevada, and in the Coast Ranges from Santa Cruz County north to Oregon. Many visitors to Yosemite Valley have seen it there on the banks of the river.

PLATE 54.   WESTERN AZALEA

44

Our western forms of LABRADOR-TEA (*Ledum glandulosum*), plate 55, are rather rigidly branched shrubs one to about five feet high with fragrant, resinous, alternate, leathery leaves. The inflorescence has few flowers in the Sierran form (var. *californicum*) and many in that from the North Coast Ranges (ssp. *columbianum*). The petals are white, about one-fourth to one-third inch long. The Sierran form inhabits boggy and wet places at 4,000 to 12,000 feet and ranges to Trinity and Modoc counties. The coastal form is from below 2,000 feet.

PLATE 55.   LABRADOR-TEA

WHITE-HEATHER (*Cassiope Mertensiana*), plate 56, is, like the preceding plant, of the Heath Family (34, 44, 75). It is shrubby, with ascending branches four to twelve inches high and has minute leaves keeled on the back. The bell-shaped flower is white to pinkish, one-fourth inch long. It grows on rocky ledges and in crevices at 7,000 to 12,000 feet and is distributed in the Sierra Nevada from Fresno County north and in the Coast Ranges from Trinity County north, to Alaska and Montana.

PLATE 56.   WHITE-HEATHER

WHITE PHLOX (*Phlox diffusa*), plate 57, is perennial, somewhat woody at the base, more or less prostrate, with almost needlelike leaves about one-half inch long. The flowers are mostly solitary at the ends of short leafy branches and about half an inch wide. It grows on dry slopes and flats at 3,300 to 11,500 feet, in the San Gabriel Mountains and Sierra Nevada and in the Coast Ranges from Glenn County to Oregon. Related white-flowered species may be more cushion-like, with shorter leaves, and more glandular.

PLATE 57.   WHITE PHLOX

PLATE 58.   ELDERBERRY

ELDERBERRY (*Sambucus caerulea*), plate 58, of the Honeysuckle Family, is a low shrub at higher altitudes and a taller one at lower elevations. It has pinnate opposite leaves with toothed leaflets. The very numerous small white flowers are in flat-topped clusters and form small bluish berries about one-fourth inch in diameter. This shrub occurs in open places below 10,000 feet from San Diego County northward through the Sierra Nevada and North Coast Ranges to British Columbia and Alberta. See also page 23.

PLATE 59.   RED-BERRIED ELDER

RED-BERRIED ELDER (*Sambucus microbotrys*), plate 59, is a low shrub with rank odor and cream-colored flowers in dome-shaped clusters or even more elongate. The red fruits are about one-sixth of an inch long and lack the powdery bloom of the berries of the preceding species. This Elder is common in moist places at 6,000 to 11,000 feet, in the San Bernardino Mountains, in the Sierra Nevada, and in the North Coast Ranges from the Yolla Bolly Mountains north and east to the Rocky Mountains.

PLATE 60.   VALERIAN

VALERIAN (*Valeriana capitata* var. *californica*), plate 60, is of the Valerian Family, characterized by small flowers that are often saccate or spurred on one side. Valerian is a perennial from strong-scented underground parts and with paired leaves that may be entire or divided. The calyx is split distally, especially as it matures, into twelve to seventeen bristles. The species is found in moist or dryer places at 5,500 to 10,500 feet from Tulare and Trinity counties north to Oregon.

PEARLY EVERLASTING (*Anaphalis margaritacea*), plate 61, of the Sunflower Family (see pages 24, 51, 57, 79, 90, 102–108) is a white-woolly perennial with slender running rootstocks. The stems are leafy, unbranched, erect, with sessile leaves mostly one to four inches long and often greener above than beneath. The flower heads are pearly white. It grows in openings in the woods, on talus and the like, below 8,500 feet, in the San Bernardino Mountains, in the Coast Ranges from Monterey County north, and in the Sierra Nevada, then to Alaska, the Atlantic Coast, and Eurasia.

PLATE 61.   PEARLY EVERLASTING

IRIS (see page 52) as represented by the species *L. missouriensis*, plate 62, is one of the most widely spread species in western North America, ranging from the mountains of San Diego County, at scattered stations in meadows and moist flats, to British Columbia, South Dakota, and Coahuila. It is a typical Iris of the familiar type with pale lilac sepals and petals to paler with lilac-purple veins. In California it grows largely between 3,000 and 11,000 feet and blooms in May and June.

PLATE 62.   IRIS

In the Buttercup Family (see pages 40, 53, 66, 83) one of the largest groups is LARKSPUR and one of the physically largest western species is *Delphinium glaucum*, plate 63. It is a coarse-stemmed leafy plant three to eight feet high with leaves three to seven inches across and divided into broad incised and toothed segments. The racemes are four to twelve inches long, many-flowered, light to dark violet-purple. Found in wet meadows and near streams at 5,000 to 10,600 feet, this species occurs in the San Bernardino and San Gabriel mountains, in the Sierra Nevada, and thence to Alaska and the Rocky Mountains.

PLATE 63.   LARKSPUR

PLATE 64.   WESTERN MONKSHOOD

Another widespread species of the Buttercup Family is WESTERN MONKSHOOD (*Aconitum columbianum*), plate 64, a highly variable complex with many local forms. With rather tuberous roots, it has mostly erect and stout stems two to several feet tall and well distributed palmately lobed leaves. The flowers are in rather a loose inflorescence, are over half an inch long, and are usually purplish-blue with the upper sepal arched and called a hood. Monkshood grows in moist places like willow thickets, at 4,000 to 8,000 feet, from the Sierra Nevada and northernmost Coast Ranges northward.

PLATE 65.   LUPINE

To the Pea Family (see pages 17, 71, 84, 98) belong the lupines, one of California's prides. One LUPINE (*Lupinus Breweri*), plate 65, is a low matted perennial with silvery-silky foliage and stems one to two inches high. The leaves are crowded and have seven to ten leaflets less than an inch long. Considered as a whole, since the species has a number of forms, it is found on dry stony slopes and benches between 4,000 and 12,000 feet, from the San Bernardino Mountains to Oregon. The small violet flowers are less than half an inch long.

PLATE 66.   LUPINE

Another LUPINE is *Lupinus excubitus*, plate 66, its forms from the pine belt being quite to scarcely woody at the base, rather silvery-silky, one to two feet high, and with leaflets an inch or more long. The rather fragrant flowers are blue to violet, a half inch or more long. It is common on dry slopes and in rocky places at 4,000 to 8,500 feet, from the Tehachapi Mountains to Lower California and along the desert base of the Sierra Nevada.

BLUE FLAX (*Linum perenne* ssp. *Lewisii*), plate 67, is perennial, usually with several leafy stems one-half to two and one-half feet high and with numerous narrow leaves. A true Flax, it has very fibrous and tough tissues in the stem. The blue flowers have five petals about one-half inch long, which fall away easily when the plant is disturbed. It grows on dry slopes and ridges, generally between 4,000 and 11,000 feet, from Lower California to Alaska and eastern Canada and Texas.

PLATE 67. BLUE FLAX

California-Lilac (see pages 42 and 72) is here represented by SQUAW CARPET (*Ceanothus prostratus*), plate 68, a prostrate shrub with its branches rooting and forming mats a yard or more across. The leaves often have three sharp apical teeth and are light green in color. The blue flowers are in small inflorescences. It inhabits open flats in pine forests at 3,000 to 6,500 feet, from Calaveras and Alpine counties north to the North Coast Ranges and to Washington.

PLATE 68. SQUAW CARPET

ALPINE GENTIAN (*Gentiana Newberryi*), plate 69, is of the Gentian Family (see pages 75, 85). It is a perennial with one-flowered stems two to four inches high and with broadly spatulate leaves one to two inches long. The broadly funnelform corolla usually has dark bands without and is light green with greenish dots inside. It occurs in moist meadows and on banks, mostly at 7,000 to 12,000 feet, from Tulare to Siskiyou counties and to southern Oregon. Flowering is from July to September.

PLATE 69. ALPINE GENTIAN

49

PLATE 70.   SKY PILOT

In the Phlox Family (see pages 19, 35, 45, 76) is the genus *Polemonium* which is here illustrated for *P. eximium*, SKY PILOT, by plate 70. It is a viscid perennial four to twelve inches high, from a woody base, and with a strong musky odor. The leaves are divided into numerous 3- to 5-parted leaflets. The flowers are crowded into a head and are narrow-funnelform to cylindrical, about half an inch long, and deep blue in color. The species grows on dry rocky ridges and slopes at 10,000 to 14,000 feet, from Tuolumne County to Tulare and Inyo counties.

PLATE 71.   PHACELIA

In the Waterleaf Family the large genus PHACELIA has many forms (see pages 20, 77, 86). A low annual type is represented by *Phacelia curvipes*, plate 71, branched and diffuse, to about four inches high. The leaves are largely basal and entire, although some related species have them few-lobed. The broadly bell-shaped corolla is almost one-third of an inch long and bluish to violet with a white throat. It is a species of dry slopes at 3,500 to 8,000 feet, from the mountains of San Diego County to the east slope of the Sierra Nevada and to Utah.

PLATE 72.   SELFHEAL

Our native SELFHEAL (*Prunella vulgaris* ssp. *lanceolata*), plate 72, is in the Mint Family (see pages 21, 88). A low perennial herb, it has leaves one to two inches long and stems four to twelve inches high. The flowers are in a dense group, violet to purplish, one-half inch or longer, two-lipped. It is found in moist woods and about ditches, at below 7,500 feet, in most of montane California. It ranges to Alaska and the Atlantic states.

Also of the Mint Family (pages 21, 22, 50, 88) are the sages, the most common mountain SAGE being *Salvia pachyphylla*, plate 73. It is a somewhat sprawling subshrub with grayish leaves one to one and one-half inches long and with dense spikes of flowers subtended by purplish bracts. The large blue to violet-blue corolla is almost an inch long, narrow, two-lipped, and with well exserted stamens. This Sage frequents dry rocky places at between 5,000 and 10,000 feet, from the San Bernardino Mountains to Lower California.

PLATE 73.   SAGE

PENSTEMON or BEARD TONGUE (see pages 35, 36) is represented at higher elevations by *P. heterodoxus,* plate 74, a slender-stemmed perennial four to ten inches high and with the leaves largely basal. The inflorescence has two to four rather distinct many-flowered clusters. The blue-purple corolla is about one-half of an inch long, narrow, and with equal upper and lower lips. It is found on rocky slopes and in alpine meadows at from 8,000 to 12,000 feet and ranges from Tulare County to Plumas County in the Sierra Nevada.

PLATE 74.   PENSTEMON, BEARD TONGUE

In the Sunflower Family (see pages 24, 47, 52, 57, 79, 90, 102–108) there are several small asterlike plants, one of which is an ASTER (*A. alpigenus* ssp. *Andersonii*), plate 75. It grows in meadows at 4,000 to 11,500 feet, from Tulare County to southern Oregon and is found also in the San Jacinto Mountains of Riverside County. Its grasslike leaves are basal and tufted, while the flower heads are solitary and showy, sometimes an inch or more in diameter.

PLATE 75.   ASTER

51

PLATE 76. ERIGERON

Another dwarf asterlike plant is ERIG-ERON (*pygmaeus*), plate 76, cushionlike in habit, from a woody taproot, and more or less hairy throughout. The leaves are in a dense rosette and the heads solitary on erect leafless stems one to two inches high. The involucral bracts are purplish or black-purple and the outer flowers (rays) are purple to lavender and about one-fourth inch long. This Daisy occurs on rocky slopes and flats at 10,000 to 12,000 feet, from the Mount Whitney region north to Mount Rose, Nevada.

PLATE 77. LEOPARD LILY

LEOPARD LILY or PANTHER LILY (*Lilium pardalinum*), plate 77, is one of California's more conspicuous lilies (see pages 62, 93). The stout stems attain a height of from three to eight feet and bear three to four whorls of nine to fifteen leaves as well as some scattered ones. The nodding flower has anthers about one-half inch long and the yellow, spotted perianth-segments are recurved to the middle or below. This Lily forms large colonies on stream banks and in springy places at below 6,000 feet, in the Palomar Mountains, Sierra Nevada, and Coast Ranges from Santa Barbara County to Humboldt County.

PLATE 78. IRIS

An IRIS (see page 47) that is not blue in many of its forms is *Iris Hartwegii*, plate 78, of wooded slopes at 2,000 to 6,000 feet, from Butte and Plumas counties to Kern County. The rather narrow petals vary from deep yellow and lavender to pale yellow and cream and are one and one-half to over two inches long. Largely in dry woods in the San Bernardino Mountains is a purplish to bluish-violet form called ssp. *australis*.

52

A yellow-flowered Wild Buckwheat (see pages 15, 30, 64, 95) is SULPHUR FLOWER (*Eriogonum umbellatum*), plate 79, with several named variants in California. It is perennial from a woody caudex, usually with several low stems leafy at the tips and bearing umbels of simple or branched rays, which may or may not bear bracts near their middle. The flowers are in terminal subcapitate to open clusters and range from yellow to orange or reddish. Commonly it has bright yellow flowers with a reddish tinge and grows on dry and rocky places below 11,000 feet in most of the California mountains and to the Rocky Mountains.

PLATE 79. SULPHUR FLOWER

Of the yellow-flowered species of BUTTERCUP one of the most common in our pine belt is *Ranunculus alismifolius* var. *alismellus,* plate 80. The erect stems are one-half to one foot high and have simple entire leaves. The petals are about one-fourth of an inch long, five in number. It is often abundant in meadows and on wet banks, at 4,500 to 12,000 feet, in the San Jacinto and San Bernardino mountains, the Sierra Nevada, the Coast Ranges from Glenn County north, and to Washington and Montana.

PLATE 80. BUTTERCUP

Another *Ranunculus* is *R. Eschscholtzii* (ALPINE BUTTERCUP), plate 81. In a variety of forms it occurs about rocks and in meadows between 8,000 and 13,500 feet in the San Jacinto and San Bernardino mountains, in the Sierra Nevada, and from Trinity and Siskiyou counties to Alaska and the Rocky Mountains. The stems are to about six inches long and the leaves have rounded lobes. The petals are to about one-half inch long. See also page 40.

PLATE 81. ALPINE BUTTERCUP

PLATE 82. WALLFLOWER

WALLFLOWER (*Erysimum perenne*), plate 82, belongs to that large assemblage, the Mustard Family with its four petals and ovary up inside the flower (see pages 16, 67, 95, 96). It is a short-lived perennial, four to twelve inches high, with the root-crown clothed with the remains of the old leaves. The yellow petals are over half an inch long and the spreading seed pods two to three inches. It is found in dry places at 7,000 to 12,000 feet and is distinctive of the area between Tulare County and Mount Shasta. It is also in the Scott Mountains in Trinity County and in the Yolla Bolly Mountains.

PLATE 83. CUSHION-CRESS

CUSHION-CRESS (*Draba Lemmonii*), plate 83, also of the Mustard Family, has a spreading leafy cushion with leaves less than an inch long and leafless flowering stems to four inches high. The yellow petals are about one-fourth of an inch long and the oval seed pods one-fourth to slightly less than one-half inch in length. The species occurs on gravelly and stony slopes, in crevices, and on talus, at 8,500 to 13,000 feet, from Tulare County to Tuolumne and Mono counties.

PLATE 84. CALIFORNIA PITCHER-PLANT

The Pitcher-Plant Family has insectivorous plants with leaves modified for catching insects and with flowers nodding at the ends of longish stems. CALIFORNIA PITCHER-PLANT (*Darlingtonia californica*), plate 84, has each conspicuously veined leaf ending in a hooded tip with two appendages. The sepals are yellow-green and the petals purple, an inch or more long. It occurs in marshy and boggy places at 300 to 6,000 feet, in Del Norte, Trinity, Siskiyou, Plumas, and Nevada counties and in Oregon. The flowering season is from April to June.

The Stonecrop Family (see page 31) is mostly succulent and its flower is on the plan of five. One STONECROP (*Parvisedum Congdonii*), plate 85, is only one to three inches high and diffusely branched. It occurs in rocky places below 5,000 feet in the Sierran foothills from Eldorado County to Tulare County. Its petals are one-twelfth of an inch long and always spreading, while another species, *P. pumilum*, has slightly longer petals that become erect in age. This species grows in rocky places and beds of vernal pools below 4,000 feet, from Sutter County to Merced County.

PLATE 85. STONECROP

Another STONECROP is *Sedum spathulifolium*, plate 86, a perennial with slender rootstocks and rather prominent rosettes of leaves. The flowering stems are two to twelve inches high and bear reduced leaves. The usually yellow petals are to one-third of an inch long. In one form or another this plant occurs in rocky places at 2,500 to 7,000 feet, from the San Bernardino and San Gabriel mountains to Monterey and Eldorado counties and northward to British Columbia.

PLATE 86. STONECROP

In the same family with Stonecrop is LIVE-FOREVER (*Dudleya cymosa*), plate 87, a variable, more or less glaucous plant with mostly oblanceolate fleshy leaves two to six inches long. The flowering stems are to about one foot tall and bear reduced leaves. The bright yellow to reddish petals are about one-half of an inch long. In the pine belt this Live-Forever ranges to elevations of 9,000 feet and occurs in much of our mountainous area from the San Bernardino Mountains north.

PLATE 87. LIVE-FOREVER

PLATE 88.   PODISTERA

PODISTERA (*P. nevadensis*), plate 88, is of the Carrot Family (see pages 42, 43, 73, 74). It is a bunched stemless perennial with a crown of fibrous sheaths and with tufted short leaves having three to seven divisions. The flowering stems are to about one inch long and bear headlike clusters of yellowish flowers. It is found generally above timberline at 10,000 to 13,000 feet and ranges in the Sierra Nevada from Placer County to Tuolumne and Mono counties, occurring also in the San Bernardino Mountains. Flowering is from July to September.

PLATE 89.   MONKEY-FLOWER

In the Figwort Family (see pages 35–38, 51, 88–89, 102) one of the interesting groups is the monkey-flowers. A particularly beautiful MONKEY-FLOWER of the higher mountains is *Mimulus Tilingii*, plate 89, with its bright green leaves to about one inch long and its few-flowered stems six to sixteen inches high. The yellow two-lipped flowers are an inch or more long and have brown-spotted ridges in the throat. It occurs on wet banks at 6,400 to 11,000 feet, in the San Jacinto and San Bernardino mountains, the Sierra Nevada, and to British Columbia.

PLATE 90.   MONKEY-FLOWER

Another smaller MONKEY-FLOWER is an annual, *Mimulus Whitneyi*, plate 90, remarkable for having flowers of various colors: pale yellow with maroon blotches and lines, or purple with similar paler areas. The plant is one to two inches high; the leaves less than an inch long; and the almost sessile flowers are about one-half inch long. This Monkey-Flower grows in gravelly places at 6,000 to 11,000 feet, in Fresno and Tulare counties.

In the Sunflower Family (pages 24, 47, 51, 79, 90, 102–108) and in many ways quite like the Sunflower itself is WYETHIA or MULE-EARS (*Wyethia mollis*), plate 91, a resin-dotted perennial densely woolly when young. Its stems are one to three feet high and its basal leaves eight to sixteen inches long. The stems bear one to four heads about an inch across and with rather few ray-flowers (the petallike outer ones in the head). This Wyethia occurs on dry wooded slopes and in rocky openings at 5,000 to 10,600 feet, from Fresno County to Siskiyou and Modoc counties and to Oregon. Related species are not so woolly or may have basal and stem leaves almost alike.

PLATE 91.   MULE-EARS, WYETHIA

Hulsea algida, plate 92, is another member of the Sunflower Family and is often called ALPINE GOLD. It is a glandular-pubescent, disagreeably odorous perennial with few to many one-headed, leafy stems four to sixteen inches high. The entire basal leaves are up to six inches long and the heads two or more inches across. It is found on rocky peaks at 10,000 to 14,000 feet and ranges from Mount Whitney to Mount Rose, Nevada. *Hulsea nana* is another species of the Mount Lassen and Mount Shasta area and has pinnately lobed leaves.

PLATE 92.   ALPINE GOLD, HULSEA

GOLDENROD is represented in the higher mountains by *Solidago multiradiata*, plate 93, a perennial with erect stems two to sixteen inches high. The basal leaves are one to four inches long; the heads few to many, each with about thirteen ray-flowers. It inhabits sunny rocky or grassy places at mostly 7,000 to 12,500 feet and occurs in the Sierra Nevada from Tulare County north through the Cascades to Alaska and Siberia, as well as east to Labrador.

PLATE 93.   GOLDENROD

PLATE 94.   SILVER MAT

Another genus of the Sunflower Family and one often lacking ray-flowers in its heads is *Raillardella*, illustrated here by SILVER MAT (*R. argentea*), plate 94. It is a silky-woolly perennial to four inches high, the heads bearing an involucre about half an inch long. It grows in dry rocky places at 9,000 to 12,000 feet, in the San Bernardino Mountains, the Sierra Nevada, and north to Oregon. Its gray leaves distinguish it from the Sierran *Raillardella scaposa* (page 108) with green glandular leaves and sometimes with short rays, and from the Coast Range species *R. scabrida* with branched stems.

PLATE 95.   GROUNDSEL

One of the largest groups in the Sunflower Family is GROUNDSEL, represented here by *Senecio Fremontii*, plate 95; see also page 109. Perennial from a branching base, it has slender decumbent stems and rather fleshy rounded leaves. The heads are solitary at the ends of the branches with the involucre almost one-half inch high. The yellow ray-flowers are one-fourth to one-third inch long. It grows about rocks at 8,500 to 12,400 feet, in the San Bernardino Mountains, the Sierra Nevada, and to British Columbia.

PLATE 96.   GROUNDSEL

Another montane GROUNDSEL is *Senecio werneriifolius*, plate 96. It is perennial, with tufted basal leaves white-woolly to greenish and smooth, entire or nearly so. The several stems are two to six inches high, are almost leafless, and bear one to six flower heads, each with nine to thirteen rays one-fourth to one-third inch long. This species is found in dry rocky places at 10,400 to 13,000 feet, in the Sierra Nevada from Tulare and Inyo counties to Fresno and Mono counties, and in the Rocky Mountains. Similar species occur in more moist places and are more widespread geographically.

58

FLOWERS WHITE TO PALE CREAM
OR PALE PINK OR GREENISH

Section Three

ARROWGRASS (*Triglochin maritima*), figure 43, is a densely tufted plant one to two feet high. The leaves are fleshy and the flowers small and green, in terminal bracted spikes, and have six concave perianth-segments. The fruit is a cluster of six one-seeded structures (carpels) that separate when mature. The species occurs in wet alkaline flats and boggy places, often near hot springs, below 7,500 feet and ranges in the San Bernardino Mountains, Sierra Nevada, and to Alaska, the Atlantic Coast, and Eurasia.

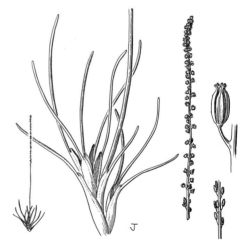

FIGURE 43. ARROWGRASS

ARROWHEAD, TULE-POTATO, or WAP-PATO (*Sagittaria latifolia*), figure 44, is one to three feet high and with narrow to broad and arrow-shaped leaf-blades. Its white flowers are in whorls of three. The staminate flowers are commonly above and the pistillate below, in an open branching inflorescence. The petals are half an inch or more long. Arrowhead is found at the edge of ponds or slow streams and in meadows, below 7,000 feet, and ranges through most of California to British Columbia and the Atlantic states. The tubers are edible.

FIGURE 44. ARROWHEAD OR TULE-POTATO

RUSH or WIREGRASS (*Juncus Parryi*), figure 45, is a tufted perennial to a foot high; the leaves are grooved at the base but terete above. The stems are slender and bear one to three mostly brownish flowers, each having six sepallike parts, three stamens, and a pistil. This Rush grows in rather dry rocky places above 9,000 feet in the San Bernardino Mountains and at from 6,000 to 12,500 feet in the Sierra Nevada, also from the Yolla Bolly Mountains north to British Columbia. Other rushes, with similar often greenish flowers and usually a more open inflorescence, are common in wet places.

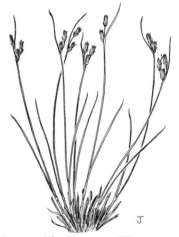

FIGURE 45. RUSH OR WIREGRASS

FIGURE 46.   WOOD RUSH

FIGURE 47.   MARIPOSA-LILY

FIGURE 48.   WASHINGTON LILY

WOOD RUSH (*Luzula parviflora*), figure 46, also of the Rush Family, but with softer, more hollow stems, is a tufted plant to two feet high. The leaves are thin, two to five on a stem; the flowers are in open compound panicles and are minute, greenish to pale brown, each with six sepallike segments. It is found in moist places in woods, at 3,500 to 11,000 feet, from Kern and Mendocino counties to Alaska, Labrador, and Eurasia. Other species have the flowers in more compact, even in spikelike, clusters.

MARIPOSA-LILY (*Calochortus invenustus*), figure 47, is a bulbiferous plant with slender erect stems to about one and one-half feet high, with linear leaves to six or eight inches long, and with white to dull lavender flowers an inch or more long. Sometimes there is a purplish spot below the gland near the base of the petal. This attractive member of the Lily Family occurs on dry brushy and grassy slopes and flats at 4,500 to 9,000 feet, from Tulare County to Tuolumne County, from Santa Clara County to Monterey County, and from the Tehachapi Mountains to the Laguna Mountains.

WASHINGTON LILY (*Lilium Washingtonianum*), figure 48, has stems four to six feet tall and light green leaves in several whorls of six to twelve. The flowers are up to twenty or more, trumpet-shaped, white with a few reddish dots. It grows among bushes on dry slopes and flats at 4,000 to 7,000 feet, from Fresno County to Siskiyou County. SHASTA LILY is a smaller form at the base of Mount Shasta and CASCADE LILY a somewhat purplish type from Humboldt and Siskiyou counties north. See pages 52, 93, 94.

ADDER'S-TONGUE or DOGTOOTH-VIOLET
(*Erythronium purpurascens*), figure 49,
also of the Lily Family (pages 13, 28, 29,
52, 83, 93, 94), is a perennial from a
deep-seated corm and has two almost
basal leaves and a flowering stem often
to about one foot high. This bears one to
eight flowers, white with yellow base or
tinged purple in age, and about one-half
inch long. Occurrence is along streams
and in meadows or on brushy or forested
slopes at 4,000 to 8,000 feet from Tulare
County to Shasta County. A golden yellow species (*E. grandiflorum*) is in the
North Coast Ranges.

FIGURE 49.   ADDER'S-TONGUE

REIN ORCHID (*Habenaria dilatata* var.
leucostachys), figure 50, has fleshy tuber-
like roots and scattered green leaves on
a stem one to five feet high. This bears
a terminal spike of white flowers, each
about half an inch long. This attractive
fragrant orchid is frequent in wet and
springy places below 11,000 feet in the
mountains from San Diego County northward to British Columbia. Flowering is
from May to August. See also page 14.

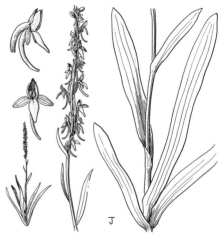

FIGURE 50.   REIN ORCHID

KNOTWEED (*Polygonum bistortoides*),
figure 51, of the Buckwheat Family
(pages 14, 15, 30, 53, 64, 95) is a perennial with several erect, slender, simple
stems one to two feet high. The leaves
are mostly near the base; the white flowers in terminal thick-cylindric spikes one-
half to two inches long. Each flower has
six perianth-segments about one-fifth
inch long. The plant is common in wet
meadows and along streams, mostly at
between 5,000 and 10,000 feet, in the
San Jacinto and San Bernardino mountains, the Sierra Nevada and North Coast
Ranges, thence to Alaska and the Atlantic Coast.

FIGURE 51.   KNOTWEED

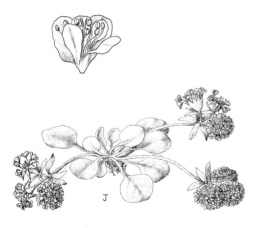

FIGURE 52.   WILD BUCKWHEAT

FIGURE 53.   WILD BUCKWHEAT

FIGURE 54.   PIGWEED, GOOSEFOOT

WILD BUCKWHEAT, an immense group in California, is here represented by *Eriogonum Lobbii,* figure 52, with a stout, woody, few-branched caudex and leaves in tufted rosettes. These are densely woolly beneath. The flowering stems (scapes) are more or less flat on the ground and bear white to rose flowers. It is found on dry rocky walls and slopes at 5,500 to 12,000 feet in the Coast Ranges from Lake County north and in the Sierra Nevada from Inyo and Mariposa counties to Plumas County. See also page 95.

Another WILD BUCKWHEAT (*Eriogonum saxatile*), figure 53, is of a somewhat different type than in figures 52 and 134 and illustrates to some extent the tremendous array of forms this remarkable group assumes in California. It too is perennial; it has closely white-felted leaves, slender ascending flowering stems, and white or tinged flowers one-fourth inch long. It is to be sought on dry rocky slopes and ridges, generally at 4,000 to 11,000 feet, from the San Jacinto Mountains to the southern Sierra Nevada and in the Santa Lucia Mountains. It flowers from May to July.

PIGWEED or GOOSEFOOT (*Chenopodium Fremontii*), figure 54, is met along trails and similar disturbed places between 5,000 and 8,500 feet, from the mountains of southern California to Mono County and British Columbia. It is an erect annual to two or three feet high, with triangular-hastate leaves greenish above and whitish beneath. In the Goosefoot Family the flowers are minute, greenish or membranous, without petals. Such plants as the Garden Beet and Spinach, as well as Desert-Holly, belong here.

The Portulaca Family, fleshy and often with only two sepals (see pages 15, 30) is here shown in Toad-Lily (*Montia Chamissoi*), figure 55, a perennial with slender runners that are more or less buried and that produce bulblets. The leaves are of several pairs and the petals white to pinkish and about one-fourth inch long. It is found in wet places in meadows and along streams at 4,000 to 11,000 feet, from San Diego County to the Sierra Nevada and Modoc County, and in the North Coast Ranges from Lake County north, then to Alaska and Minnesota. It blooms from June to August.

FIGURE 55.   TOAD-LILY

Starwort (*Stellaria longipes*), figure 56, belongs to the Pink Family, with its paired leaves and its flower on the plan of five. It is a somewhat tufted perennial from creeping rootstocks, to about eight inches high, and with narrow leaves to about one inch long. The flowers are one to few, with cleft petals one-fourth inch long. It is a dainty little plant of moist places at from 4,500 to 10,500 feet, in the San Bernardino Mountains, Sierra Nevada, and from the Yolla Bolly Mountains north to Alaska and to the Atlantic Coast.

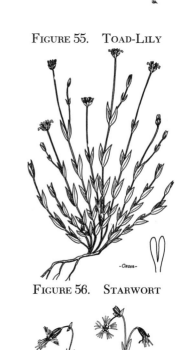

FIGURE 56.   STARWORT

Another member of the Pink Family is Catchfly or Campion (*Silene Lemmonii*), figure 57, a slender-stemmed perennial to about one foot tall, glandular in the upper parts, although not sticky as in some other species. In this *Silene* the flowers are mostly nodding and the petals cleft into four linear lobes. The sepals are grown together into a tube. It is common in open woods at 3,500 to 8,000 feet, from the Cuyamaca Mountains of San Diego County and the Santa Ynez Mountains of Santa Barbara County north to Oregon. See also page 40.

FIGURE 57.   CATCHFLY or CAMPION

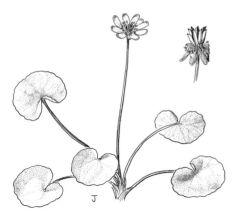

FIGURE 58.   MARSH-MARIGOLD

FIGURE 59.   COLUMBINE

FIGURE 60.   ANEMONE

MARSH-MARIGOLD (*Caltha Howellii*), figure 58, belongs to the Buttercup Family with its separate pistils. It is a fleshy perennial with large simple mostly basal leaves and solitary flowers having white sepals and no petals. It is found in marshy and boggy places at 4,500 to 10,500 feet, in most of the Sierra Nevada and in the North Coast Ranges and southern Oregon. Flowers appear from May to July.

The three plants shown on this page all belong to the Buttercup Family (see also pages 40, 47, 48, 53, 67, 83). The COLUMBINE (*Aquilegia pubescens*), figure 59, is a white to yellowish species with erect flowers. It is found at higher elevations, such as 9,000 to 12,000 feet, particularly on talus and in rocky places. It ranges from Tulare and Inyo counties on the south to Tuolumne County on the north. Each petal has a long, hollow, nectar-bearing spur and thus attracts sphinx moths which do the pollinating.

ANEMONE (*A. Drummondii*), figure 60, is another perennial. From its stout root-crown arise one to several stems to about one foot tall. They and the dissected leaves are quite hairy. The white or bluish-tinged sepals are about half an inch long; petals are lacking. Occupying places in talus and gravel or rocks at 5,000 to 10,600 feet, this Anemone ranges in the Sierra Nevada from Inyo County north and in the Coast Ranges from Trinity County north to Alaska and Alberta. It flowers from May to August.

MEADOW-RUE (*Thalictrum Fendleri*), figure 61, is a perennial to three or four feet high, with leaves two to four times divided into broad leaflets. Staminate and pistillate flowers are on separate plants, greenish, with four to seven deciduous sepals and no petals. The one-seeded pistils have three to four ribs on each side. This species is found mostly in damp places at 4,000 to 10,000 feet and ranges from the mountains of San Diego County north to Oregon and Wyoming.

FIGURE 61.   MEADOW-RUE

BITTER-CRESS (*Cardamine Breweri*), figure 62, is of the Mustard Family (see pages 16, 17, 54, 68, 95, 96). It is perennial with creeping rootstocks and attains a height of one to two feet, bearing undivided leaves or with three to five ovate parts. The white petals are about one-fourth inch long and are followed by linear seed-pods to an inch long. It grows along streams at 4,000 to 12,000 feet, in the San Bernardino Mountains and along the Sierra Nevada to the mountains of Humboldt County, thence to British Columbia and Wyoming. Flowering is from May to July.

FIGURE 62.   BITTER-CRESS

ROCK-CRESS (*Arabis Holboellii* var. *retrofracta*), figure 63, is another plant of the Mustard Family and a member of a large genus *Arabis* which has many species in America and Eurasia. This Rock-Cress has stems to over two feet high and whitish to pinkish flowers. It produces long slender almost straight reflexed pods. It is found in dry stony places at 1,800 to 5,000 feet in the North Coast Ranges and at 6,000 to 10,500 feet in the Sierra Nevada and the San Bernardino Mountains. In many related species the seed-pods are strongly arched and often spreading.

FIGURE 63.   ROCK-CRESS

FIGURE 64.   ROCK-CRESS

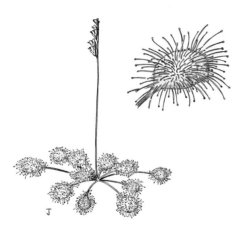

FIGURE 65.   SUNDEW

FIGURE 66.   SAXIFRAGE

Another quite different ROCK-CRESS from the one in figure 63 is the little *Arabis platysperma*, figure 64, less than one foot high and with entire leaves. The seed-pods are flat, ascending, and one to almost three inches long. It grows on dry stony flats and slopes at 5,500 to 11,200 feet, from the San Jacinto to the San Gabriel mountains and in the Sierra Nevada and North Coast Ranges from Tulare and Glenn counties north respectively.

SUNDEW (*Drosera rotundifolia*), figure 65, is an insectivorous plant of cold wet bogs and swamps below 8,000 feet, from Tulare County to Mount Shasta and from Sonoma County north. It is circumpolar. The leaves are in a spreading rosette, with their upper surfaces clothed with tentaclelike, gland-tipped, reddish hairs that bend over and entrap insects. The whitish or pinkish petals are soon shed. This genus with three others, including the famous Venus' Flytrap of North Carolina, constitute the interesting Sundew Family, known for its insectivorous habit.

The Saxifrage Family has the floral parts somewhat united into a tube that may be partly grown to the ovary (see pages 17, 31, 41). SAXIFRAGE (*Saxifraga punctata* ssp. *arguta*), figure 66, is perennial with roundish coarsely toothed leaves and flowering stems a foot or more in height. The white petals have two yellow dots at the narrowed base. It is found on moist stream banks at 6,500 to 11,200 feet in the San Bernardino Mountains, Sierra Nevada, and from the Yolla Bolly Mountains north to Washington and the Rocky Mountains.

Another SAXIFRAGE (*Saxifraga debilis*), figure 67, exemplifies quite a different type from that shown in figure 66. It forms small tufts two to four inches high. The lower leaves are reniform in outline, mostly with three to five lobes, and the upper are much reduced. The white petals are oblong-spatulate. It is found in damp shaded places about overhanging rocks at 11,000 to 12,000 feet in the Sierra Nevada and occurs also in the Rocky Mountains.

FIGURE 67.   SAXIFRAGE

Still another member of the Saxifrage Family is the odd little MITREWORT (*Mitella Breweri*), figure 68, with small greenish petals pinnately parted into linear lobes. The leaves are all basal and roundish. This species occurs on damp shaded slopes at 6,000 to 11,500 feet, in the Sierra Nevada from Tulare County north, ranging to British Columbia and Montana. Other related species with whitish petals occur in the Sierra Nevada and North Coast Ranges.

FIGURE 68.   MITREWORT

In the Rose Family (see pages 32, 41, 42, 70, 96, 97), which is related to the Saxifrage Family, are many woody plants like cherries and apples and many herbaceous species like Geum and Strawberry. The sepals and petals are united into a floral tube that surrounds the ovary or ovaries. The example here discussed, CREAM BUSH (*Holodiscus Boursieri*), figure 69, is a compact shrub two to three feet high and is found on dry rocky slopes at 4,000 to 9,600 feet. It ranges in the mountains of southern California, in the Sierra Nevada, and Coast Ranges from Lake County north. It has small, hairy, toothed leaves and small whitish to pinkish flowers.

FIGURE 69.   CREAM BUSH

FIGURE 70. HORKELIA

FIGURE 71. IVESIA

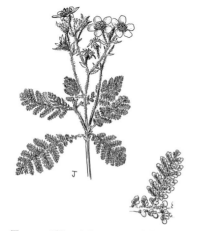

FIGURE 72. MOUNTAIN MISERY

Likewise in the Rose Family and related to the cinquefoils and potentillas is HORKELIA (*H. fusca* ssp. *parviflora*), figure 70, a slender-stemmed perennial with dark green, glandular-pubescent leaves divided into five to ten pairs of leaflets. The white spatulate petals are scarcely one-sixth of an inch long. This Horkelia grows in open places at 2,500 to 10,500 feet from Tulare to Del Norte and Siskiyou counties and to Washington and Wyoming. Closely related forms and species are common in much of California and frequently are quite aromatic.

IVESIA (*I. santolinoides*), figure 71, and the next two plants are also of the Rose Family. Ivesia is an attractive perennial herb with the silvery-silky, much dissected, vermiform leaves largely in a basal rosette and with the slender suberect stems to about one foot tall. The white petals are quite small, but the flowers are many and give a Baby's-Breath effect. It is found on dry gravelly slopes and ridges at from 5,000 to 12,000 feet and from the San Jacinto Mountains to the Sierra Nevada of Eldorado County.

A low shrubby plant is MOUNTAIN MISERY (*Chamaebatia foliolosa*), figure 72, with many leafy branches and glandular, heavy-scented young twigs that soon exfoliate. The leaves are viscid, much divided into crowded elliptical ultimate parts. The white petals are one-fourth to one-third of an inch long. It is common as a ground cover in open forests between 2,000 and 7,000 feet and ranges from Shasta County to Tulare County. The flowers appear between May and July.

A rather tall and erect shrub is BITTER CHERRY (*Prunus emarginata*), figure 73. It is deciduous in winter and grows to a height of five to twenty feet. The several white flowers in flattish clusters are about one-fourth inch long and produce bright red bitter fruits. It grows on dryish rocky ridges or dampish slopes below 9,000 feet and ranges in the mountains from southern California to British Columbia and Idaho. It flowers in April and May.

FIGURE 73.   BITTER CHERRY

The Pea Family (see pages 17, 48, 72, 84, 98) is one of the largest and most important plant families as a source of food, ornamentals, and lumber. It is noteworthy for its very large genera, like *Acacia*, for example. Another is CLOVER, one species of which, *Trifolium longipes*, figure 74, is a common perennial in moist places below 9,000 feet from the San Jacinto Mountains through the Sierra Nevada and North Coast Ranges to Washington and Idaho. It varies in different parts of its range, but has its flowers sessile in peduncled heads and has three leaflets to each leaf.

FIGURE 74.   CLOVER

Another montane CLOVER, also perennial, but with very slender stems and 1–3-flowered heads of narrow whitish florets is *Trifolium monanthum*, figure 75. It also has a number of forms, but in general is found about grassy moist places at from 5,000 to 11,500 feet and from the San Jacinto Mountains to Plumas County in the Sierra Nevada. A much larger species is *T. Wormskioldii* with coarser growth and large heads of flowers half an inch long.

FIGURE 75.   CLOVER

FIGURE 76.   LOCOWEED

FIGURE 77.   TOBACCO BRUSH

FIGURE 78.   SNOW BUSH

One of the largest genera in North America and of the Pea Family is *Astragalus,* some species of which are called LOCOWEED because they poison livestock, others RATTLEWEED, since they have dry inflated pods in which the seeds rattle about in the wind. The LOCOWEED shown in figure 76 (*Astragalus Bolanderi*), is a perennial to one and one-half feet high, each leaf having seventeen to twenty-five leaflets. The flowers are several, in loose racemes, and one-half to two-thirds of an inch long. It is found in dry stony and sandy flats and meadows at 5,200 to 10,000 feet from Tulare County to Plumas County.

Perhaps California's showiest genus of shrubs is *Ceanothus* (see pages 42, 49) with over forty species in the state. It belongs to the Buckhorn Family and has small flowers with a flattish central disk. Many species of so-called California-Lilac are beautiful, but no such claim can be made for TOBACCO BRUSH (*Ceanothus velutinus*), figure 77, yet it is a conspicuous, spreading, round-topped, evergreen shrub with dark green leaves varnished above and paler beneath. The flower-clusters are one to two inches long. It occurs on open wooded slopes at 3,500 to 10,000 feet and ranges from Tulare County to Trinity, Humboldt, and Modoc counties, then to British Columbia and South Dakota.

Another rather homely but conspicuous and common *Ceanothus* is SNOW BUSH (*C. cordulatus*), figure 78, found on dry open flats and slopes at 3,000 to 9,500 feet, from the San Jacinto Mountains north through the Sierra Nevada and in the North Coast Ranges from Lake County north to Oregon. It is low, spinose, grayish-glaucous, and intricately branched. The flowers are in dense white clusters to about one inch long.

In the Evening-Primrose Family (see pages 18, 19, 32, 73, 100) a small-flowered but common and often noticeable annual is the slender-stemmed GAYOPHYTUM (*G. diffusum*), figure 79. It has alternate, entire, narrow leaves and scattered flowers with petals to about one-sixth inch long. The ovary is beneath the flower (inferior) and the four petals are white to pinkish in age. Found in dry open places at 3,000 to 11,000 feet, it can be expected in the mountains from San Diego County north to Washington and Dakota.

FIGURE 79.   GAYOPHYTUM

Another member of the same family but with a reduced flower (two instead of four sepals, two petals, two stamens) is ENCHANTER'S-NIGHTSHADE (*Circaea alpina* var. *pacifica*), figure 80. It is rather a low perennial with a simple erect stem, a few pairs of thin petioled leaves, and minute flowers in terminal racemes. The fruit is nutlike and covered with hooked hairs. It grows in deep woods below 8,000 feet in the San Bernardino Mountains, Sierra Nevada, and Coast Ranges from Marin County north to British Columbia and to the Rocky Mountains.

FIGURE 80.   ENCHANTER'S-NIGHTSHADE

The next four species are in the Carrot Family (see pages 42, 43, 56), a large group of aromatic herbs with hollow stems, often decompounded leaves, and minute flowers in umbels (flower-clusters with the divisions arising at one level). SQUAW ROOT (*Perideridia Gairdneri*), figure 81, has fusiform tubers, slender stems, and pinnately dissected leaves. This and related species are often conspicuous in drying meadows below 11,000 feet, from San Diego County to British Columbia. The tubers were eaten by the Indians.

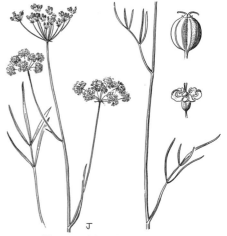

FIGURE 81.   SQUAW ROOT

FIGURE 82. SWEET-CICELY

SWEET-CICELY (*Osmorhiza chilensis*), figure 82, has the aromatic quality of the other members of the Carrot Family. Perennial, slender-stemmed, one to three feet high, it has rounded leaf-blades divided into rather broad leaflets. The flowers are greenish-white to white, minute; the fruit is cylindrical. It grows in woods below 10,000 feet from San Diego County to Alaska and the Atlantic Coast, as well as in South America.

FIGURE 83. HOG-FENNEL

HOG-FENNEL (*Lomatium macrocarpum*), figure 83, is another perennial, ten to fifteen inches high, with the leaves in a subbasal tuft and much divided into linear segments. The flowers are many, white to yellowish or even purplish, and the flat fruits are one-half to two-thirds of an inch long. The species is found in dry stony places below 8,000 feet from Kern and San Luis Obispo counties north to British Columbia.

FIGURE 84. ANGELICA

Another member of the Carrot Family is ANGELICA (*A. Breweri*), figure 84, a conspicuous plant three to four feet high and with large lance-shaped leaflets as divisions of the large leaves. The inflorescence is also large and has twenty-two to forty-five unequal primary divisions or rays and oblong to oval fruits. This Angelica is found on open wooded slopes at 3,000 to 8,600 feet in the Sierra Nevada from Inyo County to Shasta County. Another montane species is *A. linearifolia* with linear leaflets.

With over forty species in California
and forming a conspicuous portion of
our woody vegetation is the genus
Arctostaphylos or Manzanita (Heath
Family, see pages 34, 44, 45), characterized by shrubby habit, smoothish red
stems, and small white or pinkish urn-
shaped flowers. PINEMAT MANZANITA
(*Arctostaphylos nevadensis*), figure 85,
is sprawling or prostrate, with intricately
branched stems and light green leaves.
It inhabits moist places or dry rocky
slopes in woods at 5,000 to 10,000 feet,
from Tulare County north in the Sierra
Nevada and from Lake County north in
the Coast Ranges, to Washington.

FIGURE 85.   PINEMAT MANZANITA

Another member of the Heath Family
is the WESTERN BLUEBERRY (*Vaccinium
occidentale*), figure 86, a low shrub
with thin leaves less than an inch long.
The fruit is a blue-black berry with a
bloom, sweetish in taste, but rather inferior as blueberries go. The species occurs in wet places at 5,000 to 11,000 feet
in the Sierra Nevada from Tulare County
north, to Modoc and Trinity counties,
thence to British Columbia and the
Rocky Mountains.

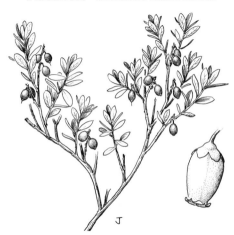

FIGURE 86.   WESTERN BLUEBERRY

In the Gentian Family (pages 49 and
85) our conspicuous montane plants have
blue flowers, but in the genus *Frasera* is
GREEN GENTIAN (*F. speciosa*), figure
87, with an open greenish-white corolla
dotted with purple and an inch or more
across. It is a coarse plant, three to six
feet high, with whorls of three to seven
leaves, and with four-parted flowers. It
is found in dry to damp places at 5,000
to 9,800 feet in the Sierra Nevada from
Fresno County north and in the Coast
Ranges from Lake County north, ranging to Washington and the Rocky Mountains. Other related green gentians occur
in most parts of the state.

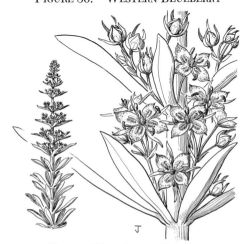

FIGURE 87.   GREEN GENTIAN

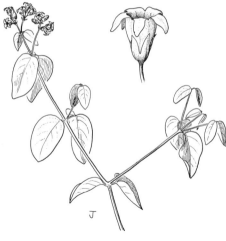

FIGURE 88.   DOGBANE

FIGURE 89.   MORNING GLORY

FIGURE 90.   WHITE-GLOBE GILIA

The Dogbane Family is well known in cultivation by such plants as Oleander and Natal Plum. In our mountains is DOGBANE (*Apocynum androsaemifolium*), figure 88, a smooth diffusely branched perennial with drooping leaves paler beneath than above. The flowers are bell-shaped, with white lobes having pinkish veins. The fruit usually consists of two ripened pods (or better, follicles) which are pendulous at maturity and two to four inches long. This Dogbane occurs in dry places at 5,000 to 9,500 feet in the San Bernardino and San Jacinto mountains and in a somewhat different form northward. A related montane species is *A. pumilum* with more cylindrical corollas.

To many Californians the MORNING GLORY is exemplified by a weedy pest, the introduced Bindweed, but some of the native species as *Convolvulus malacophyllus,* figure 89, are more attractive. This is a low vine with grayish-woolly leaves, short trailing stems, and flowers an inch or more long. It is found on dry steep slopes at 3,000 to 7,500 feet, from Siskiyou and Trinity counties to the Sierra Nevada into Tulare County. Related plants occur also in southern California.

In the Phlox Family (see pages 19, 20, 35, 45, 50, 77, 86) there is here presented WHITE-GLOBE GILIA (*Ipomopsis congesta* ssp. *montana*), figure 90, a matted perennial with palmately lobed leaves. The small white flowers are crowded into heads and have the five lobes to the corolla and three to the style characteristic in the family. It is found mostly in dry places at 7,000 to 12,000 feet in the Sierra Nevada and White Mountains to eastern Oregon.

Another member of the Phlox Family is an annual GILIA or LINANTHUS (*Linanthus Harknessii*), figure 91, with slender stems to about one foot high. The paired leaves are palmately 3–5-parted into linear lobes and the filiform ultimate branchlets of the stems bear small whitish flowers. It grows in open sandy and gravelly places at 3,000 to 10,000 feet, in the Coast Ranges from Lake County north and in the Sierra Nevada from Fresno County north, to Washington. Other related species vary in rather technical characters.

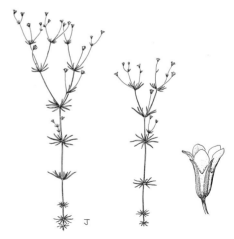

FIGURE 91.   GILIA or LINANTHUS

In the Waterleaf Family (see pages 20, 21, 50, 86, 87) there is here presented a rather coarse, often hispid perennial from a woody caudex and with stems one to two feet long. It is PHACELIA (*P. imbricata*), figure 92. The lower leaves are lobed; the flowers are crowded in dense coiled cymes arranged in open panicles. The species is extremely variable and in one form or another occurs on dry, often rocky places at 3,000 to 7,500 feet through much of California. Related species are found to above timber line.

FIGURE 92.   PHACELIA

In the same family is HESPEROCHIRON (*H. pumilus*), figure 93, a stemless perennial with leaves to about two inches long. The flowers are flat and open, to almost one inch across, densely long-hairy within. This plant occurs in moist, sometimes subalkaline places at 1,200 to 9,000 feet in the Coast Ranges from Lake County north, in the Sierra Nevada from Kern County north, and to Washington. With a more funnelform corolla and less hairy within is *H. californicus* from the San Bernardino Mountains north.

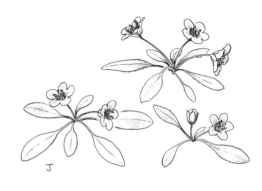

FIGURE 93.   HESPEROCHIRON

FIGURE 94.   ALPINE FORGET-ME-NOT

FIGURE 95.   WHITE FORGET-ME-NOT

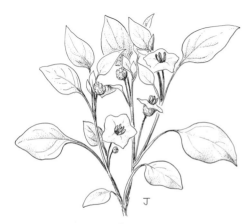

FIGURE 96.   CHAMAESARACHA

The Borage Family, another group with coiling flower-clusters called cymes, but with the ovary usually forming four one-seeded nutlets, is represented here by ALPINE FORGET-ME-NOT (*Cryptantha humilis*), figure 94, a caespitose perennial with one to several stems from a woody caudex and densely leafy. The white flower is almost half an inch broad. The species is characteristic of high dry ridges at from 6,000 to 11,400 feet, in the Sierra Nevada from Mono County to Nevada, Alpine, and Inyo counties. A similar species with smooth instead of tubercled seeds is *C. nubigena*.

Another CRYPTANTHA or WHITE FORGET-ME-NOT is *C. Torreyana*, figure 95, the common annual species of the pine belt, where it grows on dry, more or less open slopes at 1,500 to 7,500 feet, from Marin and Kern counties north to British Columbia. The stems are slender, often openly branched; flowers are small. See also page 87.

In the Nightshade Family are such familiar plants as Potato, Tobacco, Pepper, Tomato (see also page 88). Most members have quite a strong odor. In CHAMAESARACHA (*C. nana*), figure 96, we have a low perennial from tough slender underground rootstocks, with entire leaves one to two inches long. The corolla is white with five basal green spots; the fruit a berry dull white to yellowish. In the Sierra Nevada it occurs on sandy flats at 5,000 to 9,000 feet, from Mono and Sierra counties north, to Modoc and Siskiyou counties.

The Sunflower Family, with its flowers minute and crowded into heads subtended by bracts, is prominent in California (see pages 24, 47, 51, 57, 79, 90, 102–108). In the genus *Chaenactis* the florets are all tubular, the outer being enlarged as compared with the inner.

FIGURE 97.   PINCUSHION FLOWER

PINCUSHION FLOWER (*C. santolinoides*), figure 97, is a perennial with simple or branched stems to about one foot high. The leaves are mostly in basal rosettes; the cream or white heads are about half an inch high. It is found in open woods and on dry ridges at 4,500 to 8,000 feet, from the San Bernardino Mountains to the Greenhorn Mountains of Kern County. Related species are widespread in California, like *Chaenactis Douglasii*, figure 98, more or less floccose or woolly when young. The loosely woolly leaves are one to four inches long and the corollas whitish to pinkish. In somewhat varied form it ranges from 4,000 to 10,600 feet and from Lake and Tulare counties northward to British Columbia.

FIGURE 98.   PINCUSHION FLOWER

YARROW (*Achillea lanulosa*), figure 99, is another Composite or member of the Sunflower Family. In it we have a perennial quite aromatic herb with mostly finely dissected leaves and numerous small heads of white, roundish, outer petallike ray-flowers and central tubular or disk-flowers. Yarrow is found in meadows and damp places at 2,500 to 8,000 feet and, in a reduced form, to 11,300 feet, from the mountains of southern California through the Sierra Nevada and from the Yolla Bolly Mountains north to the Cascade Range and to the Rocky Mountains.

FIGURE 99.   YARROW

FIGURE 100. ADENOCAULON

FIGURE 101. PUSSYTOES

FIGURE 102. HAWKWEED

ADENOCAULON (*A. bicolor*), figure 100, a perennial from slender rootstocks, is white-woolly below and glandular in the upper parts. The leaves are near the base and white-woolly beneath. The whitish tubular flowers are in heads with the involucre formed of a single row of sub-equal green bracts. Adenocaulon is found in moist shaded woods below 6,000 feet, in the Coast Ranges from Santa Cruz County north and in the Sierra Nevada from Tulare County north, to British Columbia and Michigan.

PUSSYTOES (*Antennaria rosea*), figure 101, forms leafy gray-woolly mats from which arise the flowering stems to a few inches or almost a foot in height. The minute flowers are in woolly heads with a dry chaffy involucre that may be white or almost rose in color. It is found in dryish to moist, more or less wooded places at 4,500 to 12,000 feet and ranges in the San Bernardino Mountains, the Sierra Nevada, and from the Yolla Bolly Mountains north to Alaska and Ontario. Other species may have greenish leaves and some have the involucre brownish to dirty green.

HAWKWEED (*Hieracium albiflorum*), figure 102, is perennial, densely hairy in its lower parts, erect, with the leaves mostly basal and to six inches long. Those up the stem are reduced. The white-flowered heads are in a loose panicle, with the involucre almost half an inch long and the flowers all strap-shaped. The species occurs on dry, open, wooded slopes below 9,700 feet from San Diego County to Alaska and Colorado. See also page 110.

FLOWERS BLUE TO VIOLET

Section Four

CAMAS (*Camassia Leichtlinii* ssp. *Suksdorfii*), figure 103, is a member of the Lily Family (see pages 13, 28, 29, 39, 52, 62, 93). From a perennial bulb, it bears a basal whorl of linear leaves and a slender leafless stem one to three feet tall. The flowers are a deep blue-violet to bright blue, over one inch across. Mountain meadows at elevations of 2,000 to 8,000 feet are often sheets of this lovely plant, as far south as Napa and Tulare counties, and as far north as British Columbia. A related species (*C. Quamash* ssp. *linearis*) has one perianth-segment curved downward and more flowers in bloom on one stem at the same time.

FIGURE 103.   CAMAS

BEAVERTAIL-GRASS (*Calochortus coeruleus*), figure 104, also of the Lily Family (see page 62) is a low plant with a basal leaf about six inches long and mostly one to eight flowers. The broad bluish petals have a smooth inner surface except for the beard above the gland and the fringe on their margin. It grows in open gravelly places in woods at 3,500 to 7,500 feet from Lassen and Tehama counties to Amador County. There are several related species in northern California varying in color and technical characters.

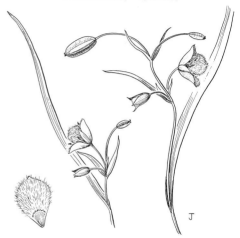

FIGURE 104.   BEAVERTAIL-GRASS

LARKSPUR (*Delphinium polycladon*), figure 105, a spurred member of the Buttercup Family, is characteristic among rocks and willows along creeks and in meadows at 7,500 to 11,150 feet, from Tulare County to Eldorado County. The flowers are dark blue to blue-purple, the sepals being about one-half inch long. The stems are several, unbranched, mostly two to three feet high, and the palmatifid leaves are largely basal. See also page 47.

FIGURE 105.   LARKSPUR

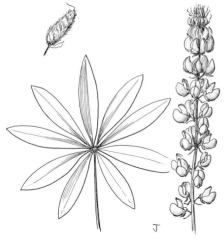

FIGURE 106.   LUPINE

A very showy LUPINE is *Lupinus polyphyllus* ssp. *superbus*, figure 106, which grows in wet places in the Sierra Nevada at elevations of 4,000 to 8,500 feet and from Tulare County to the northern borders of the state. A similar plant, ssp. *bernardinus*, is in the San Bernardino and San Jacinto mountains. The plants are two to five feet high, stout, with mostly five to nine leaflets to each leaf. The blue to purplish flowers are about one-half inch long and very numerous.

FIGURE 107   WILD PEA

Another member of the Pea Family (see pages 17, 48, 49, 71, 72, 98) is WILD PEA (*Lathyrus pauciflorus* ssp. *Brownii*), figure 107, an erect perennial with tangled stems one to two feet long and with orchid to violet-purple flowers that become bluish in age. The leaves end in simple or forked tendrils. This Pea is found on dry slopes at 4,000 to 6,000 feet, from the Tehachapi Mountains of Kern County north through the Sierra Nevada and from Tehama County into Oregon. It flowers from April to June.

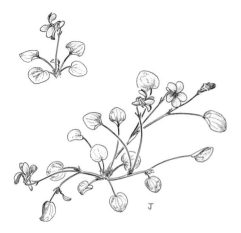

FIGURE 108.   VIOLET

California has many species of Violet (see pages 99 and 100). A VIOLET found in the mountains is *Viola adunca*, figure 108, on damp banks and at edges of meadows, from 5,000 to 11,500 feet, from San Diego County north through the Sierra Nevada and from Monterey County through the Coast Ranges, to Alaska and Quebec. The flowers are deep to pale violet in color and the spur is one-fourth to one-half of an inch long. It spreads by means of slender rootstocks.

I always associate the name Gentian
with mountain meadows as far as California is concerned (see also page 49).
One of our more conspicuous species is
EXPLORER'S GENTIAN (*Gentiana caly-
cosa*), figure 109, a perennial from a
root-crown and with thick cordlike roots.
The leaves are broad and each of the
several stems bears one to three deep
blue, funnelform-campanulate flowers an
inch or more long and with five corolla-
lobes. Found in places like meadows and
stream banks at 4,000 feet in northern
California and up to 10,500 feet in the
Sierra Nevada, this Gentian grows as far
north as British Columbia and Montana.

FIGURE 109.   EXPLORER'S GENTIAN

The SIERRA GENTIAN (*Gentiana holo-
petala*), figure 110, is an annual with
small slender roots and one flower on
each of several stems. The blue funnelform corolla is one to two inches long
and is four-lobed. It grows in wet meadows at 6,000 to 11,000 feet in the San
Bernardino Mountains and the Sierra Nevada from Tulare County to Tuolumne
County. Another montane annual is
HIKER'S GENTIAN (*Gentiana simplex*)
with a single stem that ends in a solitary
four-lobed corolla.

FIGURE 110.   SIERRA GENTIAN

Another Gentian is FELWORT (*Swertia
perennis*), figure 111, a perennial with
a short rootstock and a single erect unbranched stem to one foot high. The
leaves are mostly basal, the lower to
about four inches long. The flowers are
in a terminal panicle or raceme, mostly
five-lobed, and with a greenish or bluish-
purple tinge and almost half an inch
long. It is found in meadows and damp
places at 8,500 to 10,500 feet, from Tulare
County to Mariposa County and from
Oregon to Alaska, the Rocky Mountains,
and Eurasia.

FIGURE 111.   FELWORT

FIGURE 112.   JACOB'S LADDER

FIGURE 113.   WATERLEAF

FIGURE 114.   DRAPERIA

In the Phlox Family (see pages 19, 20, 35, 45, 50, 77) we have JACOB'S LADDER (*Polemonium caeruleum* ssp. *amygdalinum*), figure 112, a perennial from a rootstock and with solitary erect stems one to three feet long. The leaves have nineteen to twenty-seven leaflets and the blue flowers are spreading–bell-shaped, an inch across. It is found in wet places at 3,000 to 11,000 feet, in the San Bernardino Mountains and from Tulare to Siskiyou and Modoc counties, thence to Alaska and the Rocky Mountains.

In the Waterleaf Family (pages 20, 50, 77, 87) is the WATERLEAF itself (*Hydrophyllum occidentale*), figure 113, with an elongate rhizome and hairy stems to two feet high, although often much lower. The bell-shaped corolla is bluish to whitish, about one-third inch in length. It occurs on dryish or moist, somewhat shaded slopes at 2,500 to 8,000 feet, in the Tehachapi Mountains of Kern County, in the Sierra Nevada, and in the Coast Ranges from Monterey County north to Oregon and Idaho. Flowers appear from May to July.

Another member of the Waterleaf Family is DRAPERIA (*D. systyla*), figure 114, a low diffuse perennial with slender stems and a branched root-crown. The leaves are in pairs and one to two inches long. The corolla is pale violet, about one-half inch long. Draperia grows on dry slopes in woods at 2,400 to 8,000 feet, from Siskiyou and Trinity counties to Tulare County and flowers from May to August. It is named for J. W. Draper, an American historian.

In *Phacelia hydrophylloides,* figure
115, also of the Waterleaf Family, we
have a PHACELIA with a thick woody
taproot and several spreading to ascending stems to almost a foot long. The
leaves are distributed along the stems
and are largely one to two inches long.
The flowers are closely clustered, violet-
blue to pale, broadly bell-shaped, half
an inch across. It is occasional in dry
woods at 5,000 to 9,800 feet, from Tulare
County in the Sierra Nevada to Oregon.

FIGURE 115.   PHACELIA

LUNGWORT (*Mertensia ciliata* var.
stomatechoides), figure 116, is a member of the Borage Family (see page 78).
It is rather a coarse perennial, two to four
feet high, and with many bright green
leaves. The inflorescence is open in age,
with numerous, rather tubular, light blue
flowers one-half inch or more long. It
is a plant of moist places, usually in the
shade, from 5,900 to 10,200 feet, in the
Sierra Nevada from Tulare County north
and to Oregon and Nevada. It blooms
from May to August.

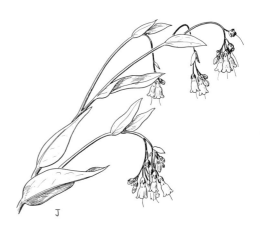

FIGURE 116.   LUNGWORT

Likewise in the Borage Family, but
with flowers more like the Forget-me-
not is STICKSEED (*Hackelia Jessicae*),
figure 117. It has few erect or ascending
stems one to two feet high from a heavy
perennial root. The leaves are well distributed along the stems which end in
open panicles of several divergent coiled
cymes with pale blue flowers about one-
sixth inch broad. The nutlets or one-
seeded fruits are prickly. This Stickseed
is common in moist places at 4,500 to
11,000 feet in the Sierra Nevada and in
the Coast Ranges from Glenn County
north, to as far as British Columbia.

FIGURE 117.   STICKSEED

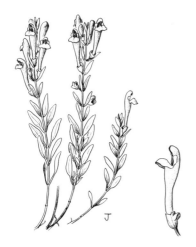

FIGURE 118.   SKULLCAP

FIGURE 119.   PURPLE NIGHTSHADE

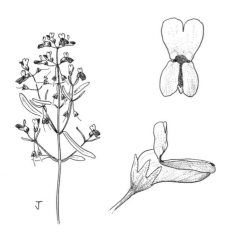

FIGURE 120.   COLLINSIA

SKULLCAP (*Scutellaria Austinae*), figure 118, is of the aromatic Mint Family with its paired leaves, two-lipped corolla, and fruit of four one-seeded nutlets (see also pages 21, 22, 51). This genus has a crestlike projection on the back of the calyx, hence the common name. The species has a deep violet-blue corolla an inch long and grows in gravelly or rocky places below 7,500 feet, from the Santa Rosa and San Jacinto mountains northward through the Sierra Nevada and in the Coast Ranges from Lake County to the Siskiyou Mountains. Other species differ in size of flowers, shape of leaves, and pubescence.

PURPLE NIGHTSHADE (*Solanum Xantii* var. *montanum*), figure 119, is a perennial to a foot or more high and more or less grayish-hairy. The spreading deep violet to dark lavender corolla is up to an inch in diameter and the plant forms greenish round berries. This variety occurs in dry places, as along trails, at 5,000 to 9,000 feet, from the San Bernardino Mountains to the Sierra Nevada of Nevada County. At lower elevations woodier and much taller forms of the species occur.

COLLINSIA (*C. Torreyi*), figure 120, of the Figwort Family (pages 22, 35–38, 51, 56, 89, 102) is remarkable in having the middle lobe of the lower lip of the corolla boatlike and enclosing the style and stamens. This species is erect, widely branched, to about six inches high. The corolla is one-third of an inch long, with a broadly rounded basal pouch, a pale upper lip with purple dots, and a longer deeper blue lower lip. In one form or another it is found below 11,000 feet from the San Bernardino and San Gabriel mountains to Siskiyou County.

SPEEDWELL (*Veronica alpina* var. *alterniflora*), figure 121, of the Figwort Family, is a perennial with erect stems four to twelve inches high and with four to seven pairs of leaves on each stem. The racemes are about four inches long, with spreading, bluish, four-lobed corollas one-fourth of an inch in diameter. It is found in wet places, generally between 7,000 and 11,500 feet, from Tulare County to British Columbia and Wyoming. It flowers from June to August.

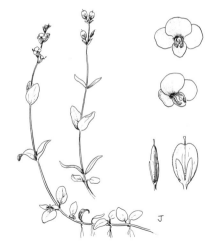

FIGURE 121.   SPEEDWELL

CALIFORNIA HAREBELL (*Campanula prenanthoides*), figure 122, a perennial with slender rootstocks, has slender erect stems one to two feet high. The flowers are generally in scattered clusters of two to five, the corolla being bright blue and almost one-half inch long. It grows in dryish wooded places below 6,000 feet, in the Coast Ranges from Monterey County north, in the Sierra Nevada from Tulare County north and to southern Oregon. Flowers appear from June to September.

FIGURE 122.   CALIFORNIA HAREBELL

PORTERELLA (*P. carnosula*), figure 123, is an erect annual, branched, a few inches high. The corolla is blue with a yellow or whitish eye, strongly two-lipped, and one-fourth to one-third inch long. It occurs in wet places, often in masses, between 5,000 and 10,000 feet, from Tulare to Lassen counties and to Oregon and Wyoming. It differs from *Downingia* of the same family in having pedicelled flowers (each on a short stem) instead of sessile, in which respect it is very near to *Lobelia*.

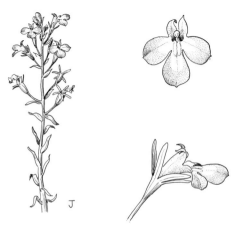

FIGURE 123.   PORTERELLA

FIGURE 124.   ASTER

FIGURE 125.   ERIGERON

FIGURE 126.   MACHAERANTHERA

The three plants presented on this page are of the Sunflower Family (see also pages 47, 51, 57, 79, 102–108). The first, ASTER (*A. foliaceus*), figure 124, has its heads surrounded by an involucre of overlapping bracts. The species is erect from a creeping rootstock, is one to three feet high, and has its middle stem-leaves sessile or clasping. The petal-like ray-flowers of the heads are purple to blue or violet, one-half inch or longer. The species has several forms and occurs above 5,000 feet in damp places or on ridges, from Tuolumne County north.

Closely related to Aster, but with the involucral bracts usually not overlapping in several series is ERIGERON (*E. peregrinus* ssp. *callianthemus*), figure 125, a fibrous-rooted perennial to about two feet high and leafy throughout. Its thirty to eighty rays are mostly rose-purple and to an inch long so that it makes a very handsome Daisy, as so many plants in this family are called. It grows in meadows at 5,500 to 10,500 feet from Tulare County north to Alaska.

In the same group of genera and much like *Aster*, but with a taproot instead of rhizome or fibrous roots, is another Daisy, MACHAERANTHERA (*M. shastensis*), figure 126. Its stems are six inches to a foot high and the stem-leaves are well developed. The heads are one to many, with involucral bracts somewhat spreading-reflexed and with eight to fifteen violet rays. It is variable and occurs in dry open gravelly places at 4,000 to 11,000 feet from the Sierra Nevada north to Oregon.

FLOWERS YELLOW TO ORANGE

Section Five

FALSE-ASPHODEL (*Tofieldia glutinosa* ssp. *occidentalis*), figure 127, of the Lily Family (see pages 13, 28, 39, 52, 62, 83, 94) has a slender perennial rootstock and stems one to two feet tall with linear leaves two to eight inches long. The light yellow flowers are about one-sixth of an inch in diameter. This is another plant of boggy places and meadows below 5,500 feet, in the Coast Ranges from Sonoma County north and in the Sierra Nevada from Tulare County north, to southern Oregon. Flowers appear in July and August.

FIGURE 127.   FALSE-ASPHODEL

BRODIAEA is often put in the Amaryllis Family, since the flowers arise at one level on the stem, but by some botanists in the Lily Family. The species shown here, *Brodiaea gracilis*, figure 128, has more or less rough stems two to ten inches high and leaves four to twelve inches long. The flowers are yellow, or purplish in age, with brown midveins on the outside. It is locally rather plentiful on gravelly plains and granitic ridges at 4,000 to 9,800 feet, from Plumas County to Mono and Mariposa counties, and flowers in June and July.

FIGURE 128.   BRODIAEA

ALPINE LILY (*Lilium parvum*), figure 129, has a rhizomatous bulb and stems one to five feet tall. The leaves are light green, two to four inches long, mostly scattered along the stem, but a few are in whorls. The orange to dark red flowers are spotted maroon and have segments one and one-half inches long or a little longer. This Lily inhabits boggy places at the edge of swamps or streams, often among alders or willows, at 4,000 to 9,000 feet, throughout the Sierra Nevada. It blooms from July to September. See also pages 52, 62, 94.

FIGURE 129.   ALPINE LILY

FIGURE 130.   SIERRA LILY

SIERRA LILY (*Lilium Kelleyanum*), figure 130, also has a rhizomatous bulb. Its stems are two to six feet high, the leaves two to six inches long, largely in whorls of three to eight. The nodding fragrant flowers are orange toward the tips or yellow throughout, with minute maroon dots, the segments of the perianth being one to two inches long. This Lily is found on wet banks and in boggy places at 4,000 to 10,500 feet, from Tulare to Siskiyou and Trinity counties and bears its flowers in July and August.

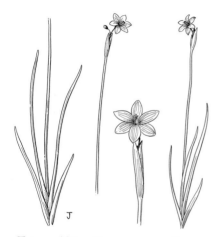

FIGURE 131.   YELLOW-EYED GRASS

In the Iris Family, with its leaves in two opposite ranks, is YELLOW-EYED GRASS (*Sisyrinchium Elmeri*), figure 131, of boggy and wet places between 4,000 and 8,500 feet, in the San Bernardino Mountains, Sierra Nevada north to Plumas County, and in Trinity County. It is a slender-stemmed plant less than eight inches tall, with narrow leaves less than half as long as the flowering stems. The orange-yellow flowers have five dark veins and are to almost an inch in diameter. They appear in July and August. See pages 47 and 52.

FIGURE 132.   PINE MISTLETOE

PINE MISTLETOE (*Arceuthobium campylopodum*), figure 132, scarcely rates as a "wildflower," but it is a curious plant that naturally attracts attention by its general yellowish or brownish color and the finder wants to know what it is. Of the Mistletoe group, but differing from ordinary Mistletoe by its minute scaly leaves and its berry on a recurved pedicel instead of being sessile, it has fragile stems that are jointed and easily break apart. It is parasitic on pines and on some forms of firs and spruces.

WILD BUCKWHEAT has many forms; see pages 15, 30, 53, 63. The one shown in figure 133 is *Eriogonum latifolium* ssp. *nudum,* mostly a few-branched perennial with basal leaves to about two inches long. The minute, clustered, yellowish to whitish or pinkish flowers are about one-tenth of an inch long and have a six-parted perianth. This subspecies is tremendously variable. It ranges from low elevations to 12,000 feet, growing in dry, somewhat rocky places from the Sierra Nevada and Coast Ranges to Washington. Related subspecies with the flowers in solitary or paired (not clustered) involucres occur in the mountains of southern California.

FIGURE 133.   WILD BUCKWHEAT

Another WILD BUCKWHEAT is *Eriogonum compositum,* figure 134, a perennial with a rather woody branched base and basal leaves white-woolly beneath. The stout flowering stems are eight to sixteen inches long and bear numerous pale yellow flowers about one-fourth of an inch long. This striking plant is found on dry rocky walls and slopes below 7,500 feet, from Lake County to Del Norte and Siskiyou counties, then to Washington and Idaho.

FIGURE 134.   WILD BUCKWHEAT

In the Mustard Family (see pages 16, 17, 54, 67, 68, 95, 96) with its acrid sap and four-merous flowers we have WINTER-CRESS (*Barbarea orthoceras*), figure 135, with rather stout stems eight to sixteen inches tall. The basal leaves are often entire, but the stem-leaves are divided. The pale yellow petals are to about one-fourth inch long, the narrow pods an inch or more. Barbarea grows on stream banks, in springy places, and in meadows, largely between 2,500 and 11,000 feet, in much of California, to Alaska and the Atlantic Coast, Asia.

FIGURE 135.   WINTER-CRESS

FIGURE 136. TANSY-MUSTARD

FIGURE 137. IVESIA

FIGURE 138. SHRUBBY CINQUEFOIL

Another crucifer (member of the Mustard Family) is TANSY-MUSTARD (*Descurainia Richardsonii* ssp. *viscosa*), figure 136, a slender pubescent annual with divided leaves. The stems are one to four feet high, usually branched above. The bright yellow petals are only an eighth of an inch long and the narrow seed-pods about one-half inch. This Mustard is to be sought in dry disturbed places between 5,000 and 11,000 feet, in the San Bernardino Mountains, the Sierra Nevada, and from Trinity County north to Washington and Alberta.

IVESIA differs from Horkelia (Rose Family, pages 32, 41, 42, 69, 70, 71) in filiform instead of dilated stamen-filaments. A yellow-flowered species is *I. Gordonii*, figure 137, with a thick woody caudex and leaves bearing ten to twenty-five leaflets which are divided to the base. The small flowers have narrow petals and one to eight pistils. This species is found in dry rocky places between 7,500 and 13,000 feet from Tuolumne and Mono counties in the Sierra Nevada and from Trinity County in the Coast Ranges to Washington and the Rocky Mountains. Very similar species occur farther south.

Potentilla, also of the Rose Family, is a large genus and is shown here by SHRUBBY CINQUEFOIL (*Potentilla fruticosa*), figure 138, a circumpolar species of woody habit, much branched, and one to four feet high. The leaves are pinnate with crowded leaflets. The yellow flowers are one to five in small groups, with round petals one-fourth to one-half inch long. It grows in moist places at 6,500 to 12,000 feet from Tulare County north and blooms from June to August.

Potentilla is represented also by *P. Drummondii*, figure 139, a perennial with fairly erect stems one to two feet high which branch above and bear few leaves. The basal leaves are pinnate, with two to five pairs of moderately spaced green leaflets. The stem-leaves have one to five leaflets. Flowers are one-half of an inch in diameter. This species is generally found in moist places at 6,000 to 13,000 feet, from Tulare and Inyo counties north and from Lake County north to British Columbia. *Potentilla Breweri* is similar, but with white-woolly leaves.

FIGURE 139.   POTENTILLA

Cinquefoil (*Potentilla gracilis*), figure 140, is another perennial but with digitate leaves, that is, all five to seven leaflets arise at one point. They are green to hairy above and quite hairy and whiter beneath. The yellow petals are almost half an inch long. It is generally found in moist places at 1,300 to 9,000 feet and grows in one form or another, since it is quite variable, from the mountains of San Diego County to Alaska and South Dakota. It blooms from May to August.

FIGURE 140.   CINQUEFOIL

Geum or Avens (*Geum macrophyllum*), figure 141, is a pinnately leaved perennial of the Rose Family characterized by the persistent hooked styles on the dry one-seeded pistils. The stems are bristly-hairy, one to three feet high; the compound leaves have very large, rounded, terminal leaflets. The petals are one-sixth of an inch or longer. It grows in moist places like meadows, at 3,500 to 10,500 feet, from the San Bernardino Mountains and Lake County north to Alaska, eastern Asia, and Labrador. It blooms from May to August.

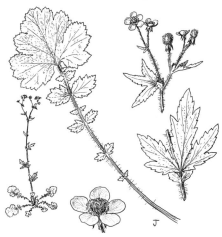

FIGURE 141.   GEUM, AVENS

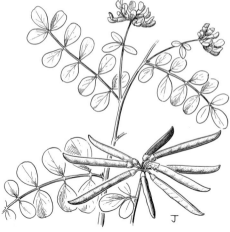

FIGURE 142. LOTUS

FIGURE 143. LOTUS

FIGURE 144. LOTUS

This page exhibits three species of LOTUS with flowers typical of the Pea Family (pages 17, 48, 71, 84). *Lotus crassifolius,* figure 142, is a perennial with stout stems one to three feet high and with pinnately compound leaves, each with seven to fifteen broad leaflets an inch or so long. The flowers are eight to fifteen in a one-sided umbel, each with a corolla about one-half inch long and forming a pod one to two or more inches long. Growing on dry banks and flats at 2,000 to 8,000 feet, the species ranges from southern California to Washington.

Quite a different LOTUS is *Lotus nevadensis,* figure 143, a prostrate perennial with many slender wiry branches forming mats one to three feet across. The leaves are appressed-hairy, each with three to five leaflets. The umbels have one to several yellow to reddish flowers one-fourth inch long. It grows on dry sandy and gravelly slopes and benches at 3,500 to 8,500 feet, in the mountains from San Diego County to Plumas County. A very similar species, *L. Douglasii,* more loosely woolly-hairy and with flowers almost half an inch long, is distributed from Lake County to Lassen County.

The third LOTUS is *Lotus oblongifolius,* figure 144, an erect or ascending perennial one to one and one-half feet high, with expanded stipules and seven to eleven leaflets to each leaf. The umbels are at the ends of long peduncles (flowerstems) and the corolla is about half an inch long. It is an inhabitant of wet places below 8,500 feet from southern California to Oregon through the Coast Ranges and the Sierra Nevada.

ST. JOHN'S WORT (*Hypericum formosum* var. *Scouleri*), figure 145, of the family bearing the same common name, has paired leaves dotted with translucent glands. With running rootstocks and erect bushy stems one to two feet high, it bears panicles of yellow flowers which are almost an inch in diameter. The stamens are many and united into three groups. It is frequent in wet meadows and on banks at 4,000 to 7,500 feet in the mountains of southern California and in the Sierra Nevada, and at lower elevations in the Coast Ranges from Monterey County north, to British Columbia and Montana.

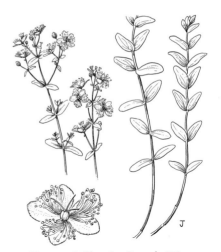

FIGURE 145.   ST. JOHN'S WORT

For Violet see pages 84 and 100. Among California's violets a good many species are yellow or yellowish, such as PINE VIOLET (*Viola lobata*), figure 146, a perennial with an erect stem to almost one foot high. The leaves are largely at the summit of the stems, palmately cleft into three to seven lobes or in some forms entire. Each flower is on a pedicel and has deep yellow petals one-third to one-half of an inch long. The two upper petals are purplish on the back. The species occurs on rather dry slopes in open woods at 1,000 to 6,500 feet, in the Cuyamaca Mountains (San Diego County), Santa Lucia Mountains, and North Coast Ranges, and in the Sierra Nevada, north to southern Oregon.

FIGURE 146.   PINE VIOLET

MOUNTAIN VIOLET is *Viola purpurea*, figure 147, a variable perennial one to several inches high. The leaves are rounded to pointed, wedge-shaped to heart-shaped at the base, and ranging greatly in size and pubescence. The petals are deep lemon-yellow, the two upper purplish-brown on the back. It is found in dry places below 11,000 feet and ranges through most of the California mountains and to Oregon.

FIGURE 147.   MOUNTAIN VIOLET

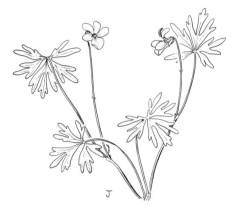

FIGURE 148.   VIOLET

FIGURE 149.   BLAZING STAR

FIGURE 150.   EVENING-PRIMROSE

Another VIOLET is *Viola Sheltonii,* figure 148, of open woods or brushy places between 2,500 and 8,000 feet in the San Gabriel and Santa Ana mountains of southern California, from the Sierra Nevada of Tulare County northward and the Coast Ranges of Santa Clara County northward, to Washington. The stems project only a little above ground and bear leaves with many linear lobes. The flowers are deep lemon-yellow with dark veins and the two upper petals are purple-brown on the back. Blooming season is from April to July.

Of the Loasa Family is BLAZING STAR (*Mentzelia congesta*), figure 149, an annual with pinnatifid leaves which are covered with close barbed hairs that make them cling to clothing. The petals are pale yellow with an orange base and scarcely one-sixth of an inch long. Stamens are many and the fruit is an elongate capsule. The species grows on dry burns and other disturbed places at 4,000 to 9,000 feet, in the Palomar Mountains, San Bernardino Mountains, and from the Sierra Nevada to Oregon.

Of the EVENING-PRIMROSE group (see pages 18, 19, 32, 73, 101) one of the most conspicuous is *Oenothera Hookeri,* figure 150. It is a variable species found in most parts of California from sea level to about 9,000 feet. The montane form, largely ssp. *angustifolia,* is mostly one to three feet high, with a reddish floral tube one to two inches long and with four broad, notched, yellow petals one to one and one-half inches long and wide. It grows in moist places at 3,000 to 9,000 feet, from the San Jacinto Mountains northward. Flowers open toward sunset and wilt the next day when the sun becomes hot.

Sun Cup (*Oenothera tanacetifolia*), figure 151, is a stemless perennial with deeply pinnatifid leaves and four-petaled yellow flowers that open in the day to almost an inch across. It is found in moist sunny places at 4,000 to 8,500 feet, from Mono and Butte counties to Modoc and Siskiyou counties and to Washington and Idaho. A closely related species, *Oenothera heterantha,* has the leaves entire or few-toothed and ranges as far south as Tulare County.

FIGURE 151.   SUN CUP

An Evening-Primrose of more local distribution is the peculiar *Oenothera xylocarpa,* figure 152, of dry benches under pines, at 7,000 to 9,800 feet and ranging from Tulare County to Mono County. It is a stemless perennial, with leaf-blades densely soft-pubescent, one to three inches long, often spotted red, pinnately parted. The vespertine flowers have petals about an inch long, bright yellow, aging red. The capsule is quite woody and four-winged.

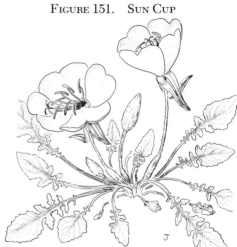

FIGURE 152.   EVENING-PRIMROSE

White-veined Shinleaf (*Pyrola picta*), figure 153, of the Wintergreen Family, is a curious little plant from a branched rootstock; its rubbery leaves are mottled or veined with white and its blades are one to three inches long. The flowers are on a leafless stem four to eight inches high and have cream to greenish petals. Some forms are almost leafless, others have little or no whitening along the veins. The plants grow in humus in dry forests at 3,000 to 10,000 feet in most of the California mountains except on the desert and range north to British Columbia. See also pages 33 and 44.

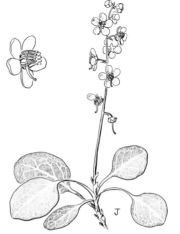

FIGURE 153.   White-veined Shinleaf

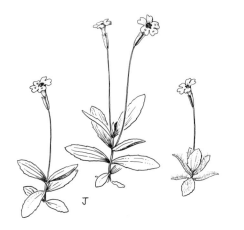

FIGURE 154.   MEADOW MONKEY-FLOWER

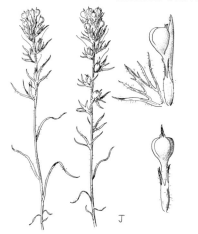

FIGURE 155.   YELLOW OWL'S-CLOVER

FIGURE 156.   BALSAMROOT

In the Figwort Family (see pages 22, 35–38, 51, 56, 88, 89) one of the large groups is Monkey-Flower or *Mimulus* and one of the most common of the montane species is MEADOW MONKEY-FLOWER (*Mimulus primuloides*), figure 154, of meadows and wet grassy banks at 3,000 to 11,000 feet. It ranges from Riverside County in southern California to Washington. Rosulate or short-stemmed, with leaves three-veined and varying from smooth to quite hairy, it bears bright yellow flowers one-half to an inch long, the corollas easily loosening and falling away.

Another member of the family is YELLOW OWL'S-CLOVER (*Orthocarpus hispidus*), figure 155, a slender-stemmed annual a few inches to a foot high, more or less hairy and with narrow leaves, the uppermost of which are 3- to 5-cleft. The flowers are about one-half inch long and borne in a terminal spike. It is found in meadows at 3,000 to 7,000 feet, rarely in San Bernardino and San Diego counties and more frequently from Glenn and Tulare counties northward to Alaska and Idaho. It flowers from May to August. Other similar species differ in technical details.

BALSAMROOT (*Balsamorhiza sagittata*), figure 156, is a member of the Sunflower Family (pages 24, 47, 51, 57, 79, 90, 103–108). It is a heavy-rooted perennial with stems one to two feet long and with large, somewhat triangular leaves. The flower heads are solitary, two to four inches across. It grows in deep, sandy, open places between 4,300 and 8,300 feet from Kern and Inyo counties to Tehama and Modoc counties and north to Canada and the Rocky Mountains. Other species may have narrower leaves.

Related to the preceding plant is HELI-
ANTHELLA (*H. californica*), figure 157,
a perennial with leafy stems one to two
feet long. The narrow leaves are five to
ten inches long and to one inch wide.
The heads are solitary, long-stemmed,
yellow in the center as well as on the
rays. Growing on grassy slopes below
7,000 feet, it ranges from Trinity County
to Shasta and Modoc counties and then
south to Kern County in the Sierra Nevada. It blooms from May to September.

FIGURE 157.   HELIANTHELLA

CALIFORNIA CONEFLOWER (*Rudbeckia
californica*), figure 158, also belongs to
the Sunflower Family. It has a leafy and
unbranched stem two to several feet high
with a single showy head on each pedun-
cle. The leaf-blades are four to ten inches
long. The center of the flower head elon-
gates into a column an inch or more high.
The species is occasional in moist places
at 5,500 to 7,800 feet from Kern County
north.

FIGURE 158.   CALIFORNIA CONEFLOWER

COMMON MADIA, also of the Sunflower
Family, is *Madia elegans,* figure 159. It
is an erect annual, branched and glandular above, hairy below, with numerous
narrow leaves. The heads have eight to
sixteen rays (petallike structures) which
are yellow or with maroon blotch at the
base. The disk-flowers are yellow or maroon. This Madia is common on rather
dry slopes and at the edge of meadows
between 3,000 and 8,000 feet through
much of California. The typical form has
rays one-fourth to one-half inch long,
while ssp. *Wheeleri* has them about one-
sixth inch long.

FIGURE 159.   COMMON MADIA

FIGURE 160.   MADIA, TARWEED

FIGURE 161.   WHITNEYA

FIGURE 162.   HULSEA

Another MADIA or TARWEED is *Madia Bolanderi*, figure 160, a perennial with woody rootstocks, a simple stem two to four feet high, and with many stalked glands in its upper parts. The leaves are linear, hairy, often four to eight inches long. The heads are few, in an open inflorescence, each with eight to twelve ray-flowers. This Madia grows in damp mountain meadows or along streamways at 3,500 to 6,700 feet, from Tulare County to the Marble Mountains and Trinity Summit, thence to Oregon. It blooms from July to September.

WHITNEYA (*W. dealbata*), figure 161, was named for John D. Whitney, to whom Mount Whitney, California's highest peak, is dedicated. The plant is perennial with one to few erect stems to about one foot long and with whitish, closely woolly, three-nerved leaves two to four inches long. The heads are solitary or few, long-stemmed, with yellow ray-flowers almost an inch long. In light, moist soil of open hillsides and of forests, the plant is uncommon, occurring at 4,000 to 7,000 feet from Fresno County to Shasta County and blooming from June to August.

HULSEA is here represented by *Hulsea vestita*, figure 162, a perennial with dense rosettes of white-woolly, entire or toothed leaves and with several stems, each ending in a single head. The ray-flowers are linear, yellow tinged with red or purple, and about one-third inch long. Sometimes reduced to a dwarf form (var. *pygmaea*) at higher altitudes, it grows in sandy or gravelly soils at 6,000 to 11,000 feet in the Sierra Nevada from Mono County south and from Frazier Mountain to the San Jacinto Range.

Another HULSEA is *H. heterochroma*,

figure 163, a robust perennial with several erect leafy stems, sticky-hairy and heavy-scented. The oblong leaves are toothed and up to four inches long. The ray-flowers are yellow to reddish-purple, twenty-five to sixty in number. This Hulsea is found at 3,000 to 8,000 feet in forest openings from the San Jacinto Mountains northward to Santa Clara County in the Coast Ranges and to Alpine County in the Sierra Nevada. It blooms from June to August.

SNAKEWEED is another member of the Sunflower Family with the individual flowers minute and in heads. *Helenium Bigelovii,* figure 164, has a stem two to four feet high, branched above or throughout, the lower leaves dead before flowering or persisting. The stem leaves are sessile and decurrent, that is, running down the stem below their insertion. The flower heads are long-stemmed, with a somewhat rounded, yellow, central disk and with thirteen to thirty showy rays to almost an inch long. It is common in moist meadowlike places from 3,000 to 10,000 feet in the mountains of southern California, the Sierra Nevada, and the North Coast Ranges, to Oregon.

GOLDEN-YARROW (*Eriophyllum lanatum*), figure 165, also of the Sunflower Family, is one of the most variable species in California and with many named varieties. It is in general perennial, erect or decumbent from a woody base, one to two feet high, with a persistent or partially deciduous woolly covering. The leaves are usually toothed or divided. The flower heads are solitary or in open groups, each with eight to thirteen yellow rays. It is rather common in brushy places below 13,000 feet from southern California to Washington and Montana.

FIGURE 163.   HULSEA

FIGURE 164.   SNAKEWEED

FIGURE 165.   GOLDEN-YARROW

FIGURE 166. BAHIA

FIGURE 167. GOLDEN-ASTER

FIGURE 168. WHITESTEM GOLDENBUSH

BAHIA (*B. dissecta*), figure 166, another plant in the Sunflower Family, is a biennial or short-lived perennial, mostly one to two feet high and with divided leaves one to three inches long. The ray-flowers are yellow and ten to thirteen in number. The species grows on gravelly open slopes and dry rocky ridges at 6,000 to 8,600 feet in the San Bernardino and Santa Rosa mountains and to Wyoming and Texas. It blooms in August and September.

Haplopappus of the Sunflower Family and related to *Aster*, is a large California genus. GOLDEN-ASTER (*Haplopappus apargioides*), figure 167, has several stems two to ten inches long and basal tufted leaves one to four inches long. The heads are usually solitary, with thirteen to thirty-four ray-flowers one-eighth to one-third inch long. Golden-Aster is found in open rocky places and about meadows at 7,500 to 12,000 feet from Tulare County to Plumas County. It blooms from July to September.

Another *Haplopappus* of quite a different type is WHITESTEM GOLDENBUSH (*H. Macronema*), figure 168. It is an undershrub four to fifteen inches high, with the numerous erect twigs of the current season white-woolly. The many leaves are to one inch long. The heads have no ray-flowers but are yellow with tubular flowers only. It grows on rocky, mostly open slopes, often on talus above timber line, between 9,000 and 12,000 feet, from Tulare County to Nevada County and to Oregon and the Rocky Mountains. It flowers from July to September.

A HAPLOPAPPUS that is also a compact
shrub is *H. Bloomeri*, figure 169, with a
woody trunk to an inch thick. The numerous leaves vary from almost filiform
or threadlike to one-eighth inch wide.
The heads are few to many in a cluster
and have only one to five ray-flowers or
sometimes none. This Haplopappus is
found in sandy or rocky places between
3,500 and 9,500 feet from Tulare County
to Siskiyou and Modoc counties and to
Washington. It is in bloom from July to
October.

FIGURE 169.   HAPLOPAPPUS

A fourth species of HAPLOPAPPUS is *H.
cuneatus*, figure 170, of slopes and cliffs,
often crevices in granitic rocks, at 2,500
to 9,000 feet and from the mountains of
southern California to Plumas and San
Luis Obispo counties. It is a deep green,
spreading, much-branched shrub with
crowded balsamic-resinous leaves, which
may be wedge-shaped to roundish (particularly in southern California) and are
one-fourth to three-fourths of an inch
long. The flower heads are in compact
clusters with one to five, sometimes no,
ray-flowers.

FIGURE 170.   HAPLOPAPPUS

RABBIT BRUSH, also of the Aster tribe
in the Sunflower Family, is very common throughout our western states, especially in subalkaline places. The species shown here is *Chrysothamnus visci-
diflorus*, figure 171. With the involucral
bracts of the heads in somewhat vertical
rows and the heads small (five-flowered)
and numerous, this plant is a characteristic Rabbit Brush. The twigs are not
woolly as in some species and the rounded shrub is usually less than three feet
high. Leaves are linear. There are several
named variants occupying dry open
places between 4,000 and 11,000 feet,
from southern California to British Columbia and Montana.

FIGURE 171.   RABBIT BRUSH

FIGURE 172.   RABBIT BRUSH

Another RABBIT BRUSH is *Chrysotham-nus Parryi,* figure 172, a shrub less than two feet high and with twigs covered with feltlike wool. The flowers are in racemes or spikes. It too has a number of named variants, but in general grows on mountain sides and flats from 3,000 to 11,000 feet, from southern California to Siskiyou and Modoc counties. Like other species of *Chrysothamnus* it blooms in the late summer and the fall.

FIGURE 173.   HYMENOPAPPUS

HYMENOPAPPUS (*H. filifolius* var. *lugens*), figure 173, is another member of the Sunflower Family. It is perennial, one to two feet high, with a basal rosette of leaves which are dissected into filiform divisions. The heads are three to eight to a stem, with yellow ray-flowers about one-fifth of an inch long. The plant is found on dry slopes at 4,000 to 7,500 feet in the mountains of southern California and to Utah and New Mexico. It blooms from June to August.

FIGURE 174.   RAILLARDELLA

RAILLARDELLA (*R. scaposa*), figure 174, is a low perennial, with its green leaves mostly basal and one to five inches long. The leafless or scapose stems are two to fourteen inches high, glandular, and each bears a head with or without ray-flowers. It grows in dry stony places and at the edge of meadows, at 6,500 to 11,000 feet, in the Sierra Nevada and to Oregon, flowering in July and August. See also page 58.

Arnica is a somewhat aromatic or glandular genus of perennial herbs from a rhizome or caudex and with mostly paired leaves. The rather large heads are solitary to several. The species shown in figure 175, *Arnica Chamissonis* ssp. *foliosa*, is one to two or more feet high, somewhat hairy, with leaves two to eight inches long. The ray-flowers are one-half to three-fourths of an inch long. It is common in meadows and moist places at 5,000 to 11,000 feet, in the San Bernardino Mountains and the Sierra Nevada, thence to Alaska and the Rocky Mountains.

FIGURE 175.   ARNICA

One of the large genera in the Sunflower Family is *Senecio* with common names like Groundsel and Ragwort. In it the involucre of the head is made up largely of equal bracts with much shorter ones at their base. One of the common montane species is *Senecio triangularis*, figure 176, of wet meadows and stream banks at 4,000 to 11,150 feet, in the San Jacinto, San Bernardino, and San Gabriel mountains, the Sierra Nevada, and the Coast Ranges from Trinity County north to Alaska and to the Rocky Mountains. The elongate-triangular leaves and flat-topped inflorescence with heads bearing six to twelve ray-flowers are characteristic.

FIGURE 176.   GROUNDSEL, RAGWORT

Quite a different Groundsel is *Senecio pauciflorus*, figure 177, with a stem six to twelve inches high and the leaves chiefly basal. They are one-half to one and one-half inches long, rounded or often somewhat lobed. The heads are usually several, somewhat reddish to orange, and with or without ray-flowers. It grows in wet meadows at 6,000 to 11,000 feet, from the Sierra Nevada to Alaska and Labrador.

FIGURE 177.   GROUNDSEL

FIGURE 178.   NOTHOCALAIS

NOTHOCALAIS is a dandelionlike member of the Sunflower Family with all its flowers in a head strap-shaped or raylike. *Nothocalais alpestris*, figure 178, is a stemless perennial with a basal rosette of mostly toothed or pinnatifid leaves and solitary heads at the tips of their peduncles. It is found in meadows and on open slopes at 7,000 to 11,500 feet, from the Sierra Nevada to Modoc and Siskiyou counties, then north to Washington. It blooms in July and August.

FIGURE 179.   HAWKWEED

HAWKWEED (*Hieracium horridum*), figure 179, also has all its flowers strap-shaped or ligulate. It is perennial, with few to several shaggy-hairy stems four to fourteen inches high and with spatulate to oblong leaves. The inflorescence is open, with many small heads, each bearing about fifteen bright yellow florets. It is common in dry, more or less rocky places at 5,000 to 11,000 feet, from the Santa Rosa and San Jacinto mountains, north through the Sierra Nevada and to Oregon, blooming in July and August. See also page 80.

FIGURE 180.   HAWKSBEARD

HAWKSBEARD (*Crepis occidentalis*), figure 180, is a foot or less tall, with a close, gray, woolly covering. The leaves are four to twelve inches long, toothed to lobed, and gradually reduced up the stems. These are one to three in number, several-branched, each main branch with ten to thirty heads. The strap-shaped corollas are almost one inch long. The species, in several different variants, occurs in dry stony or rocky places at 4,000 to 9,000 feet, from Kern and Ventura counties to the Sierra Nevada, Lassen County and to Washington and Wyoming.

COUNTY MAP OF CALIFORNIA

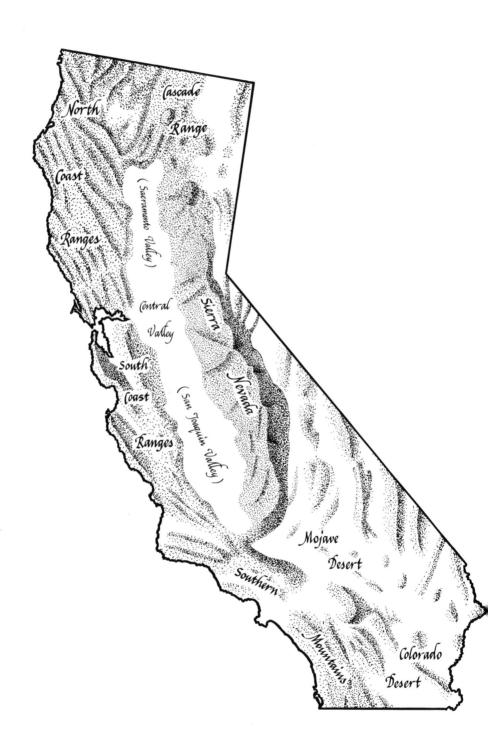

North

Cascade

Range

Coast

(Sacramento Valley)

Ranges

Central
Valley

Sierra

South

Coast

Nevada

Ranges

(San Joaquin Valley)

Mojave

Desert

Southern

Mountains

Colorado

Desert

TOPOGRAPHIC MAP OF CALIFORNIA

INDEX TO COLOR PLATES

(References are to plate numbers)

INDEX

(References are to page numbers)

115